Communications in Computer and Information Science 178

Commenced Publication in 2007
Founding and Former Series Editors:
Alfredo Cuzzocrea, Dominik Ślęzak, and Xiaokang Yang

More information about this series at http://www.springer.com/series/7899

Markus Helfert · Andreas Holzinger
Orlando Belo · Chiara Francalanci (Eds.)

Data Management Technologies and Applications

Third International Conference, DATA 2014
Vienna, Austria, August 29–31, 2014
Revised Selected Papers

 Springer

Editors

Markus Helfert
School of Computing
Dublin City University
Dublin 9
Ireland

Andreas Holzinger
Medical University Graz
Graz
Austria

Orlando Belo
Informatics
University of MInho
Braga
Portugal

Chiara Francalanci
Dipt di Elettronica e Informazione
Politecnico di Milano
Milan
Italy

ISSN 1865-0929   ISSN 1865-0937 (electronic)
Communications in Computer and Information Science
ISBN 978-3-319-25935-2   ISBN 978-3-319-25936-9 (eBook)
DOI 10.1007/978-3-319-25936-9

Library of Congress Control Number: 2015952775

Springer Cham Heidelberg New York Dordrecht London

Printed on acid-free paper

Springer International Publishing AG Switzerland is part of Springer Science+Business Media
(www.springer.com)

Preface

The present book includes extended and revised versions of a set of selected papers from the Third International Conference on Data Technologies and Applications (DATA 2014). The conference was sponsored by the Institute for Systems and Technologies of Information, Control and Communication (INSTICC) and co-organized by the Austrian Computer Society and the Vienna University of Technology – TU Wien (TUW). DATA 2014 was held in cooperation with the ACM Special Interest Group on Management Information Systems (ACM SIGMIS).

The aim of this conference series is to bring together researchers and practitioners interested in databases, data warehousing, data mining, data management, data security, and other aspects of knowledge and information systems and technologies involving advanced applications of data.

DATA 2014 received 87 paper submissions, including special sessions, from 30 countries in all continents, of which 10 % were presented as full papers. In order to evaluate each submission, a double-blind paper review was performed by members of the Program Committee.

The high quality of the DATA 2014 program was enhanced by the four keynote lectures, delivered by distinguished speakers who are renowned experts in their fields: Herbert Edelsbrunner (IST, Austria), Katharina Morik (TU Dortmund University, Germany), Matteo Golfarelli (University of Bologna, Italy), and Dimitris Karagiannis (University of Vienna, Austria).

The quality of the conference and the papers presented here stems directly from the dedicated effort of the Steering Committee and Program Committee and the INSTICC team responsible for handling all logistics. We are further indebted to the conference keynote speakers, who presented their valuable insights and visions regarding areas of hot interest to the conference. Finally, we would like to thank all authors and attendants for their contribution to the conference and the scientific community.

We hope that you will find these papers interesting and consider them a helpful reference in the future when addressing any of the aforementioned research areas. DATA 2015 will be held in Colmar, Alsace, France, during July 20 – 22.

April 2015

Andreas Holzinger
Markus Helfert
Orlando Belo
Chiara Francalanci

Preface

Organization

Conference Co-chairs

Markus Helfert   Dublin City University, Ireland
Andreas Holzinger   Medical University of Graz, Austria

Program Co-chairs

Orlando Belo   University of Minho, Portugal
Chiara Francalanci   Politecnico di Milano, Italy

Organizing Committee

Marina Carvalho	INSTICC, Portugal
Helder Coelhas	INSTICC, Portugal
Bruno Encarnação	INSTICC, Portugal
Lucia Gomes	INSTICC, Portugal
Rúben Gonçalves	INSTICC, Portugal
Ana Guerreiro	INSTICC, Portugal
André Lista	INSTICC, Portugal
Andreia Moita	INSTICC, Portugal
Vitor Pedrosa	INSTICC, Portugal
Cátia Pires	INSTICC, Portugal
Carolina Ribeiro	INSTICC, Portugal
João Ribeiro	INSTICC, Portugal
Susana Ribeiro	INSTICC, Portugal
Sara Santiago	INSTICC, Portugal
Mara Silva	INSTICC, Portugal
José Varela	INSTICC, Portugal
Pedro Varela	INSTICC, Portugal

Program Committee

James Abello	Rutgers, The State University of New Jersey, USA
Muhammad Abulaish	Jamia Millia Islamia, India
Hamideh Afsarmanesh	University of Amsterdam, The Netherlands
Markus Aleksy	ABB Corporate Research Center, Germany
Kenneth Anderson	University of Colorado, USA
Keijiro Araki	Kyushu University, Japan
Fevzi Belli	University of Paderborn, Germany
Jorge Bernardino	Polytechnic Institute of Coimbra - ISEC, Portugal

Omar Boussaid	Eric Laboratory, University of Lyon 2, France
Francesco Buccafurri	University of Reggio Calabria, Italy
Dumitru Burdescu	University of Craiova, Romania
Kasim Candan	Arizona State University, USA
Cinzia Cappiello	Politecnico di Milano, Italy
Krzysztof Cetnarowicz	AGH - University of Science and Technology, Poland
Kung Chen	National Chengchi University, Taiwan
Chia-Chu Chiang	University of Arkansas at Little Rock, USA
Christine Collet	Grenoble Institute of Technology, France
Stefan Conrad	Heinrich-Heine University Düsseldorf, Germany
Agostino Cortesi	Università Ca' Foscari di Venezia, Italy
Theodore Dalamagas	Athena Research Center, Greece
Ayhan Demiriz	Sakarya University, Turkey
Steven Demurjian	University of Connecticut, USA
Prasad Deshpande	IBM, India
Stefan Dessloch	Kaiserslautern University of Technology, Germany
Zhiming Ding	Institute of Software, Chinese Academy of Science, China
María J. Domínguez-Alda	Universidad de Alcalá, Spain
Dejing Dou	University of Oregon, USA
Habiba Drias	USTHB, LRIA, Algeria
Fabien Duchateau	Université Lyon 1/LIRIS, France
Juan C. Dueñas	Universidad Politécnica de Madrid, Spain
Tapio Elomaa	Tampere University of Technology, Finland
Mohamed Y. Eltabakh	Worcester Polytechnic Institute, USA
Barry Floyd	California Polytechnic State University, USA
Rita Francese	Università degli Studi di Salerno, Italy
Irini Fundulaki	Foundation for Research and Technology, Greece
Johann Gamper	Free University of Bozen-Bolzano, Italy
Faiez Gargouri	Miracl Laboratory, Tunisia
Nikolaos Georgantas	Inria, France
Paola Giannini	Università del Piemonte Orientale, Italy
J. Paul Gibson	TSP - Telecom SudParis, France
Boris Glavic	Illinois Institute of Technology Chicago, USA
Matteo Golfarelli	University of Bologna, Italy
Cesar Gonzalez-Perez	Institute of Heritage Sciences (Incipit), Spanish National Research Council (CSIC), Spain
Amit Goyal	Yahoo, USA
Janis Grabis	Riga Technical University, Latvia
Le Gruenwald	University of Oklahoma, USA
Martin Grund	University of Fribourg, Switzerland
Jerzy Grzymala-Busse	University of Kansas, USA
Giovanna Guerrini	Università di Genova, Italy
Amarnath Gupta	University of California San Diego, USA
Barbara Hammer	Bielefeld University, Germany
Slimane Hammoudi	ESEO, MODESTE, France

Jose Luis Arciniegas Herrera	Universidad del Cauca, Colombia
Melanie Herschel	Université Paris Sud/Inria Saclay, France
Jose R. Hilera	University of Alcala, Spain
Andreas Holzinger	Medical University Graz, Austria
Jang-eui Hong	Chungbuk National University, Republic of Korea
Andreas Hotho	University of Würzburg, Germany
Tsan-sheng Hsu	Institute of Information Science, Academia Sinica, Taiwan
Jeong-Hyon Hwang	State University of New York – Albany, USA
Ivan Ivanov	SUNY Empire State College, USA
Lifeng Jia	University of Illinois at Chicago, USA
Cheqing Jin	East China Normal University, China
Konstantinos Kalpakis	University of Maryland Baltimore County, USA
Nikos Karacapilidis	University of Patras and CTI, Greece
Dimitris Karagiannis	University of Vienna, Austria
Kristian Kersting	TU Dortmund University and Fraunhofer IAIS, Germany
Maurice van Keulen	University of Twente, The Netherlands
Jeffrey W. Koch	Tarrant County College Northeast Campus, USA
Mieczyslaw Kokar	Northeastern University, USA
Konstantin Läufer	Loyola University Chicago, USA
Sangkyun Lee	TU Dortmund, Germany
Wolfgang Lehner	Technische Universität Dresden, Germany
Raimondas Lencevicius	Nuance Communications, USA
Ziyu Lin	Xiamen University, China
Leszek Maciaszek	Wroclaw University of Economics, Poland and Macquarie University, Sydney, Australia
Florent Masseglia	Inria, France
Fabio Mercorio	University of Milano-Bicocca, Italy
Marian Cristian Mihaescu	University of Craiova, Romania
Dimitris Mitrakos	Aristotle University of Thessaloniki, Greece
Stefano Montanelli	Università degli Studi di Milano, Italy
Mirella M. Moro	Federal University of Minas Gerais (UFMG), Brazil
Erich Neuhold	University of Vienna, Austria
Boris Novikov	Saint Petersburg University, Russian Federation
George Papastefanatos	RC Athena, Greece
José R. Paramá	Universidade da Coruña, Spain
Jeffrey Parsons	Memorial University of Newfoundland, Canada
Barbara Pernici	Politecnico di Milano, Italy
Ilia Petrov	Reutlingen University, Germany
Nirvana Popescu	University Politehnica of Bucharest, Romania
Elisa Quintarelli	Politecnico di Milano, Italy
Christoph Quix	RWTH Aachen University, Germany
Alexander Rasin	DePaul University, USA
Claudio de la Riva	University of Oviedo, Spain
Colette Rolland	Université Paris 1 Panthéon-Sorbonne, France

Gustavo Rossi	Lifia, Argentina
Sudeepa Roy	University of Washington, USA
Salvatore Ruggieri	University of Pisa, Italy
Thomas Runkler	Siemens AG, Germany
Gunter Saake	Institute of Technical and Business Information Systems, Germany
Dimitris Sacharidis	IMIS Athena, Greece
Manuel Filipe Santos	University of Minho, Portugal
Maria Luisa Sapino	Università di Torino, Italy
M. Saravanan	Ericsson India Global Services Pvt. Ltd., India
Damian Serrano	University of Grenoble - LIG, France
Lijun Shan	National Digital Switching System Engineering and Technological Research Center, China
Jie Shao	National University of Singapore, Singapore
Lidan Shou	Zhejiang University, China
Alkis Simitsis	HP Labs, USA
Krishnamoorthy Sivakumar	Washington State University, USA
Harvey Siy	University of Nebraska at Omaha, USA
Yeong-tae Song	Towson University, USA
Peter Stanchev	Kettering University, USA
Manolis Terrovitis	Institute for the Management of Information Systems Athena, Greece
Babis Theodoulidis	University of Manchester, UK
Jörg Unbehauen	University of Leipizg, Germany
Robert Viseur	UMons - Polytech, Belgium
Fan Wang	Microsoft, USA
Leandro Krug Wives	Universidade Federal do Rio Grande do Sul, Brazil
Alexander Woehrer	Vienna Science and Technology Fund, Austria
Ouri Wolfson	University of Illinois at Chicago, USA
Robert Wrembel	Poznan University of Technology, Poland
Yun Xiong	Fudan University, China
Bin Xu	Tsinghua University, China
Amrapali Zaveri	Universität Leipzig, Germany
Filip Zavoral	Charles University Prague, Czech Republic
Jiakui Zhao	State Grid Electric Power Research Institute, China
Hong Zhu	Oxford Brookes University, UK
Yangyong Zhu	Fudan University, China

Auxiliary Reviewers

George Alexiou	RC Athena, Greece
Nikos Bikakis	National Technical University of Athens, Greece
Roberto Boselli	University of Milan Bicocca, Italy
Mirko Cesarini	University of Milan Bicocca, Italy
Marwa Djeffal	USTHB, Algeria

Giorgos Giannopoulos	Institute for the Management of Information Systems, Greece
Ilias Kanellos	IMIS, the Institute for the Management of Information Systems, Greece
Shigeru Kusakabe	Kyushu University, Japan
George Papadakis	IMIS, Athena Research Center, Greece
Thanasis Vergoulis	Institute for the Management of Information Systems, RC Athena, Greece

Invited Speakers

Herbert Edelsbrunner	IST, Austria
Katharina Morik	TU Dortmund University, Germany
Matteo Golfarelli	University of Bologna, Italy
Dimitris Karagiannis	University of Vienna, Austria

Contents

Using Information Visualization to Support Open Data Integration

Paulo Carvalho[1]([✉]), Patrik Hitzelberger[1], Benoît Otjacques[1], Fatma Bouali[2], and Gilles Venturini[2]

[1] Gabriel Lippmann Public Research Center, 41, rue du Brill,
4422 Belvaux, Luxembourg
{paulo.carvalho,patrik.hitzelberger,benoit.otjacques}@lippmann.lu
http://www.lippmann.lu
[2] University François Rabelais of Tours, Tours, France
fatma.bouali@univ-lille2.fr, gilles.venturini@univ-tours.fr
http://www.univ-tours.fr

Abstract. Data integration has always been a major problem in computer sciences. The more heterogeneous, large and distributed the data sources become, the more difficult the data integration process is. Nowadays, more and more information is being made available on the Web. This is especially the case in the Open Data (OD) movement. Large quantities of datasets are published and accessible. Besides size, heterogeneity grows as well: Datasets exist e.g. in different formats and shapes (tabular files, plain-text files and so on). The ability to efficiently interpret and integrate such datasets is of paramount importance for their potential users. Information Visualization may be an important tool to support this OD integration process. This article presents problems which can be encountered in the data integration process, and, more specifically, in the OD integration process. It also describes how Information Visualization can support OD integration process and make it more effective, friendlier, and faster.

Keywords: Data integration · Information Visualization · Open data

1 Introduction

The main aim of Data Integration is the selection and combination of information from different data sources/systems/entities into a unified view, in a way that users can exploit and analyze the result conveniently. For several years, Data Integration has been a major subject of study in computer science [1]. The topic has recently gained new importance due to the appearance of numerous new information sources, like Social Media, Blogs, Scientific Data, commercial data, Big Data and Open Data (OD). These data sources increase the data volumes and the potential number of providers significantly, with data coming from public and private entities, as well as from individuals. The relatively recent concept of OD is a major example of this phenomenon. OD makes information

© Springer International Publishing Switzerland 2015
M. Helfert et al. (Eds.): DATA 2014, CCIS 178, pp. 1–15, 2015.
DOI: 10.1007/978-3-319-25936-9_1

formerly hidden "inside" public and private organizations available and accessible to everyone at little or no cost and without permission limitations. In order to benefit from the presumptive high potential business-value of OD, data must be made usable, meaningful and exploitable to permit its integration [7]. This paper addresses this problem, discussing the main problems related to Data Integration with a special emphasis on the difficulties directly linked to the Integration of OD. Information Visualization – also known as InfoVis – is presented as a core and powerful approach for backing the integration process.

2   OD Integration

2.1   General Overview

The appearance of new information sources not only contributes to the growing amount of information, it also increases the heterogeneity of data sources. Data Integration processes have become even more complicated and demanding. OD integration is currently a subject of major importance. As an example, it is in the focus of the current EU research framework program Horizon 2020 [31]. One topic of the ICT2014-1 call [32] is ICT-15-2014: "Big Data and Open Data Innovation and take-up". It focuses on the entire value chains and reuse of Open (and Big) Data. This is a major challenge because of the difficulty of integrating heterogeneous datasets. Datasets may be built using completely different methods and standards (formats, schema, metadata, etc.) [2].

Today, organizations already integrate their internal data, using e.g. central repositories, data warehouses, or more process-oriented approaches like service-oriented architectures for their operational systems (Fig. 1).

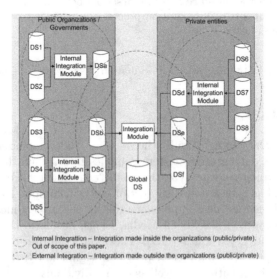

Fig. 1. Public and private data integration.

The integration of external data into these existing IT landscapes is difficult. An internal integration occurs "inside" an organization. The probability that an organization can control the format rules, policies and standards is higher than it is for external, autonomous data sources. In the following sections, we discuss OD "external integration" problems and issues, and the current status and solutions.

2.2   An Overview of the Current State

Many public organizations, from local to national and trans-national levels, have already made their data accessible on the Web. Several initiatives and directives influence these events: the European Union PSI Directive [33], Freedom of Information (FOI) initiatives in different countries and continents [34,35], Canada's Action Plan on Open Government [36], etc. One argument for publishing public data is the fact that it has been paid by the citizens in general [3]. So, over the last couple of years, the public sector has already created Open Government Data portals to open and share its data. These data portals or catalogues offer a single location where governmental data can be found [26]. Others are currently in development.

2.3   Problems and Challenges

The OD movement has not only benefits. The fact that public and private entities provide their datasets brings issues of privacy, ownership, availability, usability, accuracy and data combination [8]. Different challenges related to interoperability remain unresolved. Entities continue to build and furnish datasets without applying common standards and using heterogeneous systems. These datasets may be constructed using different and inconsistent techniques. Actually, and in general, Open Government Data initiatives publish their data using one of the following two general approaches [27]:

- The data is available on the Web as downloadable files in different formats, e.g. Excel, CSV, XML, etc.
- The data is available on the Web using RESTful APIs and SPARQL interfaces, as linked data.

Individual datasets, especially if they are complemented by metadata, are interesting and useful on their own. Nevertheless, the positive and collaborative effect of using Open Data may be higher if data of different types (scientific, social media, etc.) and delivered from several entities is combined, compared and linked. Some of the major problems and constraints in multi-source data integration are related to the following topics:

- Structure and formats used - Given the high number of different sources and datasets, it is not astonishing that Public Sector Information (PSI) is published following different modelling models (e.g. tabular, relational) [9] and formats: ZIP, CSV, XML, EXCEL, PDF, etc. Sometimes, data is even provided in non-machine-readable and/or proprietary formats;

- Metadata - Metadata is of paramount importance for data integration. A metadata schema is one of the main parts of a PSI system which should be characterized in a unified way [10]. In other words, metadata may be defined as necessary and adequate in order to understand, share and reuse data [11]. If metadata provided with a given dataset is not well-formed and/or complete, final users may have difficulties finding its related dataset [4]. Metadata provides the means to discover datasets, access them and understand them. Metadata normally refers to information about context and content (for example, a title, a description, an author, etc.) of datasets. Most of metadata schemas implemented in the public sector have been designed for national requirements. In Australia, for example, the AGLS Metadata Standard was adopted [37], New Zealand adopted the New Zealand Government Locator Service (NZGLS) while the United Kingdom chose another option, the e-Government Metadata Standard (eGMS) [28];
- Accessibility, Availability and Timeliness - If OD is commercially exploited, the providers should respond to typical business requirements in terms of accessibility, availability and timeliness. Outdated information, missing information, or information that is not accessible because of technical or other issues, cannot be the basis for reliable business processes. On the other hand, it seems unrealistic to hope that the public sector with its limited resources can offer the same service levels as commercial data providers. The integration processes must tackle these problems – or at least make them visible when they occur [12];
- Trust and Data Provenance - More and more, the need for having information and knowledge about data provenance is important. Data provenance, if it can be determined, may be used by users/data consumers to evaluate and interpret the information provided [13]. OD Integration processes and applications should be aware of data provenance, and offer efficient and reliable ways to visualize and judge it;
- Multilingualism and cultural differences - The example of the European Union, with its 28 member states and a total of 24 official languages, shows that the wealth of data that has been described above is actually a linguistic mess. Furthermore, cultural differences can already lead to different semantics for basic integration problems: An address in France is not necessarily the same "concept" as it is in Germany, for example. Ideally, information represented in different languages should not hinder its integration [14].

Even in a scenario where OD integration is technically possible, organizational and legal barriers may exclude or complicate collaboration and the sharing of data. Public and private entities may have some constraints in opening and sharing their information. They may claim ownership and/or control over certain datasets [15].

2.4   Current OD Integration Solutions

Interoperability and standards are essential to provide a solution that is able to analyse and process datasets from various data sources, using different technologies and methods. There exist several solutions and systems that provide technical and modelling frameworks for OD data providers that offer solutions that can help to achieve the goal of easy-to-integrate and standardized platforms like, e.g., Linked Open Data, CKAN, DKAN. They are presented below.

Linked Open Data. Linked Data is not a technical platform, but refers to best practices for publishing and connecting data on the web that are machine-readable and may come from different sources [16]. The adoption of these practices leads to a concept where there is a global web space in which both documents and data - from different and multiple domains - are linked. When OD is Linked Data, it is called Linked Open Data. The main objective of Linked Open Data is to help the Web of Data to identify datasets that are available under open licenses (OD sets), convert them to a Resource Description Framework (RDF) applying Linked Data principles and finally publish them on the Internet. Furthermore, Linked Open Data is concerned with two important viewpoints: data publication and data consumption [17]. Linked Open Data has more advantages and less limitations and constraints than OD. Currently, the so-called Linked Open Data cloud already provides access to information covering a large set of domains like economy, media, government, life sciences, etc. The value and potential of using all available data is huge.

In addition, while the idea behind OD is built on the concept of a social web, the notion of Linked Data is based on the semantic web approach [17] - a movement which promotes the use of common standards on the Web, encourages the inclusion of semantic data in web pages and allows data to be shared and reused by any kind of application in a cost-efficient way. Tim Berners-Lee created a five-star model which describes the different categories going from OD to Linked Open Data [29], to help and encourage entities to link their data (Table 1):

In the field of data management, Linked Open Data is gaining importance. Several Open Government Data portals, in various sectors and areas, are already using Linked Open Data principles in their systems (e.g. the Government Linked Data (GLD) Working Group [38]; the Linking Open Government Data (LOGD) Project [39]; the LOD2 project [40]).

Table 1. Sir Tim Berners-Lee five stars model.

*	Information is available on the Web under an open licence and in any format
**	(*) + Same as (*) + as structured data
***	(**) + Same as (**) + only non-proprietary formats are used (e.g. CSV instead of XLS)
****	(***) + Same as (***) + use of URI (Uniform Resource Identifier) identification - people can point to individual data
*****	(****) + Same as (****) + data is linked to other data so context is preserved - Interlinking between data

CKAN. The Comprehensive Knowledge Archive Network (CKAN) is another project related to the OD integration topic [41]. It is a web-based Open Source data portal platform for data management that provides necessary tools to any OD publisher. CKAN provides an extensive support for Linked Data and RDF. CKAN is already used by some important data catalogues worldwide (e.g. the official Open Data portal of the UK Government [42]; the prototype of a pan-European data catalogue [43]; and the Open Data Catalogue of Berlin [44]).

DKAN. DKAN is a Drupal-based[1] Open Data platform with a full suite of cataloguing, publishing and visualization features that help and support governments, non-profits organizations and universities in easily publishing data to the public [45]. Most of the core open data features that exist on CKAN are replicated in DKAN. DKAN can be appropriate for organizations already working with Drupal. In that case, deployments and maintenance may be easier.

3   Information Visualization as an Asset

Information Visualization can be extremely helpful when large amounts of data are involved. In many scenarios, users do not have the technical experience and knowledge to understand the meaning of data and how to formulate queries for the desired results. They should nevertheless be capable to discover how to link data and how data is enabled to build queries which yield the expected results [18]. Information Visualization could be a major asset to help and support end-users in these tasks. Many problems and difficulties of interpreting, filtering and viewing information can be avoided, minimized or eliminated by using Information Visualization. A Visualization System may be seen as a block which receives data as input and interacts with other entities to produce a graphical representation of the received information [19]. The strength and power of Information Visualization is the ability to present information in many and different forms, graphs and shapes (e.g.: Pie charts, Ellimaps - use nested ellipses of various sizes to build graphics [5], Treemaps, Geographical Treemaps, etc.). Depending on the purpose and meaning of the processed data, one specific chart may be easier to read and understand than another one. The following architecture is presented in order to understand how visualization may be an advantage in the way information is selected, viewed and obtained (Fig. 2).

In the solution presented above, an Information Visualization block is used in the Integration Module as a component in the integration process. It provides a way to visually present the dataset information and apply filters to them in a visual form. Based on these facts, and because OD deals with different and heterogeneous data sources and multiple types of data, we argue that Information Visualization can ease the manipulation, understanding and integration process of the data that is generated and the data that is provided by new information

---

[1] Drupal is a Content Management System which has grown in popularity in the last few years due to its openness, modularity and features [30].

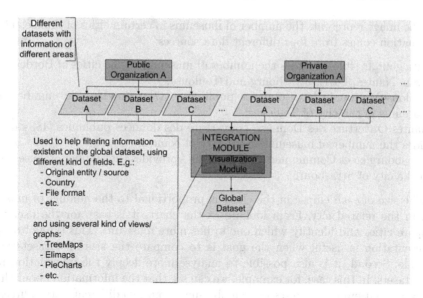

Fig. 2. Data integration with visualization.

sources. Information Visualization can be used to analyse and understand raw data and metadata in both internal and external integrations. Problems and difficulties in understanding, filtering and viewing information can be avoided or minimized by applying this paradigm. An example of how Information Visualization could help a user to quickly visualize external integration issues is presented in Fig. 3:

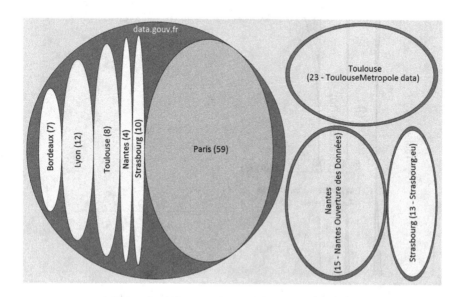

Fig. 3. Ellimap representing museum information.

The image represents the number of museums in certain cities of France. The information comes from four different data sources:

- data.gouv.fr [46] - provides the number of museums in the cities of Bordeaux, Lyon, Nantes, Paris, Strasbourg and Toulouse;
- ToulouseMetropole.data [données publiques] [47] - provides the number of museums in the city of Toulouse;
- Nantes Ouverture des Données - ouverture des données publiques [48] - provides the number of museums in the city of Nantes;
- Strasbourg.eu et Communauté Urbaine [49] - provides the number of museums in the city of Strasbourg.

The size of each ellipse in the chart is proportional to the number of museums in the related city. From looking at the chart, it is easy for the user to compare cities and identify which one(s) has more museums. This type of data representation is useful when the goal is to compare the size of datasets for example. Second, it is also possible to analyse more deeply the information in the datasets. In this case, for example, we can see that the information about the number of museums in Nantes can be obtained from two different data sources: *data.gouv.fr* and *Nantes Ouverture des Données*. This is a typical scenario where specific information can be obtained from different sources. For example, if a user wants to know the number of museums in Nantes, by analysing quickly only visual information, the user can easily notice the existence of a data incoherence. *data.gouv.fr* indicates that there are 4 museums in Nantes while *Nantes Ouverture des Données* claims the existence of 15 museums. A user will have to choose one data source for this information, which means he or she has to determine which data source is the more reliable, based e.g. on information like which

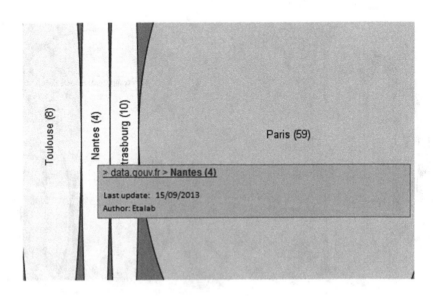

Fig. 4. Ellimap used to visualize dataset's metadata.

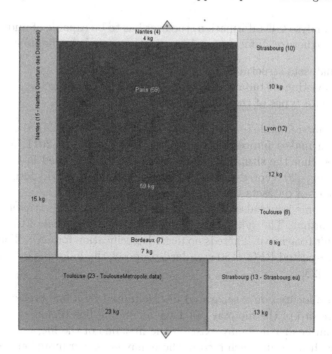

Fig. 5. Geographical Weighted Map representing museum information.

one has been modified/updated more recently, etc. An Ellimap can be used for this purpose. It is possible to visualize the metadata of datasets (e.g. by the use of tooltips over a given Ellimap). The example below demonstrates how the user may see additional information to help her or him to analyse and identify the required information (e.g. identify the more reliable data source compared to another one) (Fig. 4).

Another kind of chart could be used for the same purpose (display the information related with the museums in France) complementing the information with the location of the analysed data sources: the geographical-weighted Map. In Fig. 5, the same example as presented above is shown. The difference is that the information is organized into rectangles which are positioned according to the location of the data sources (e.g. Nantes' OGD source information is displayed on the North-West side of the graph – corresponding to the geographical location of Nantes in France; data from Strasbourg is shown on the North-East side of the graph, etc.).

3.1 Interpreting OD

Not all organizations use the same standards to publish their datasets - if standards at all are used. This is one of the major problems concerning OD integration. Since datasets may be in different formats, with or without metadata, understanding how they are organized may be a major issue. Before being able to view and analyse OD information, the user has to be capable of understanding the datasets structure. Information Visualization may help the user to:

- describe datasets structure;
- represent existing metadata delivered along with the datasets;
- show the data types of the different data entities.

Non machine-readable (e.g. PDF) and/or proprietary (e.g. XLS) data formats limits or even makes impossible to achieve these goals. In 2011, a study has demonstrated that the situation related with the formats used in OD field was not encouraging [20]. For example, PDF was the most spread format. Only a limited number of datasets was published in open (1.8)

However, some issues also exist along the advantage of the extreme simplicity of the CSV format. The syntactic and semantic interpretation of CSV datasets can be difficult to achieve. There is no formal specification for CSV. The informal and widely respected RFC 4180 standard proposes the following syntactic rules for CSV files [25]:

- Records are located on a separated line delimited by a line break;
- The last record of the file may not have an ending line break;
- An optional header line may exist on the first line of the file;
- Within the header and each record, there may be one or more fields separated by commas;
- Each field can be enclosed in double quotes;
- Fields containing line breaks, double quotes and commas should be enclosed in double-quotes;
- In fields where double-quotes are used to enclose them, a double-quote must precede another one appearing inside a field to escape it.

Based on these facts, we have focused our work on the analysis of OD CSV datasets because:

- Without having the knowledge of the dataset structure, we are not able to understand it in order to integrate it and/or use it;
- CSV is a machine-readable format;
- CSV is a basic and non-proprietary format;
- CSV is a wide spread format in OD field;
- Semantic and syntactic interpretation of CSV files structure can be eased by the use of Information Visualization;
- There is no tabular information analysis technique (e.g. Table Lens [6]) that satisfies completely our purpose (global overview of OD CSV files structure).

After several months working on this thematic, we have developed the concept of *Piled Chart* - an Information Visualization chart that is able to present in an effective manner the structure of an OD CSV file.

Piled Chart. *Piled Chart* has been designed in order to provide a visual and global overview of a CSV file. The current results are encouraging, but still need optimizations. It is currently possible to show the global structure of a CSV file using a grid-based approach and a colour code. The main characteristics of a *Piled Chart* are:

- Rows with the same structure - rows with the same data types on the same column index - are grouped (*piled*) into a unique row;
- Columns with the same structure - columns with the same data types on the same row index - are grouped (*piled*) into a unique column;
- A colour code is used to represent the data type for every cell. For now the data types supported are limited (String, Number, Date and Percentage). The colour code may also be used to represent possible errors and/or warnings.

With this approach, the structure of a CSV file can be represented on a reduced area - because rows and columns with a similar structure are grouped (*piled*). The figure below shows part of a OD CSV file which contains information about the trees in the parks of the Versailles city [50] (Fig. 6).

This simplified example of a CSV file explains the main idea behind the *Piled Chart*. The file has a simple structure of 3 rows and 10 columns. Matching representation of this file under a *Piled Chart* is showed in the figure below (Fig. 7).

The use of a *Piled Chart* to represent the file structure will help the user to determine easily the types of the cells present in the file. The chart is composed of three columns (two *piled-columns* + one "normal" column) and two rows (one normal row + one *piled-row*). With a quick look over the figure, the user can extract following information:

entityid	nomlatin	nomfranca	numvoie	adresse	nomsite	typesite	libvoie	quartier	wgs84
02a46(	Pyrus ca	Poirier a fl∈	4	RUE DE	JARDIN	JARDIN	RUE A	JUSSIEL	48.80ᵀ
04063	Quercus	Chene des	72	72	RUE SQUARE	SQUARE	RUE D	MONTRI	48.80∢

Fig. 6. OD CSV file example - Trees existing on the parks of Versailles.

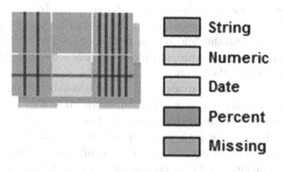

Fig. 7. Piled chart example (Color figure online).

- The cells are all composed by String or Numbers (based on a colour code);
- The first row only has String values in every cell;
- All the others rows (2 rows) have the same structure;
- The file is composed - from left to right - of 3 columns with the same structure (cells with a String value), followed by 1 column, and finally ends with a set of 6 columns with the same structure (cells with a String value).

The number of columns covered by a *piled column* may be determined by counting the regions defined by the brown lines. At the same time, the number of rows covered by a piled-row may be determined by counting the regions defined by the dark-blue lines. Missing values detection, which is an important functionality, is also straightforward. Because of the *piled*-strategy adopted, the advantage of using *Piled Chart* grows with the complexity of the analysed CSV file. However, more time must be spent to improve the current status of *Piled Chart*. E.g. actually, *Piled Chart* is only capable to show the structure of one file. It is not able to show simultaneously the structure of several files in order to compare them. In the following table, a brief summary of *Piled Chart* features is presented (Table 2):

Table 2. Piled chart features.

+	Global view of a CSV file's structure using the smallest area possible
+	Possibility to visualize each cell type and each cell value
+	Missing values detection
+	User-friendly: zoom-in/zoom-out function; tooltips with information; etc.
−	Not able to analyse 2 different files simultaneously
−	Limited to CSV files analysis

4   Conclusions and Further Work

OD offers many benefits, potential applications and services to society in general. However, it also has some constraints, barriers and issues. OD integration may be a complex task to accomplish and the related challenges and issues will continue to be an important field of research. Besides the technical problems, some entities - both in the private and public sectors – continue to be reluctant to collaborate and share their data.

Fortunately, more and more data is nevertheless being published and is already available. Having access to these massive quantities of information is however not enough to realize the above-mentioned potential. The quote of Gertrude Stein "Everybody gets so much information all day long that they lose their common sense" fairly resumes the meaning of having access to large amounts of information but being completely impotent to harness and use it because of an incapacity to interpret and analyze it.

Governments and others that wish to open their data should do it in an organized and well-structured manner, furnishing datasets accompanied by metadata describing their content. Due to the use of standards and the application of principles to publish data over the web, in the long run, Linked Open Data may be a solution to open, share and reuse data in distributed environments in an effective and cost-efficient way, so that it can be made available and accessed by any kind of application.

OD will continue to be a rapidly-evolving and heterogeneous data source. Information Visualization can be a powerful tool for supporting the OD integration process. Its methods and means may be used to provide mechanisms to analyze and process large datasets rapidly and efficiently, in both internal and external integration scenarios, giving a visual overview of the dataset structures and helping the user to understand its content, detect possible errors in datasets and data incoherencies, and show metadata so it can be used for filtering, etc.

Based on our current research, we intend to build a Visualization platform to support complex OD integration, trying to make the whole process more effective, more intuitive and quicker. To reach this objective, the platform will use advanced and innovative types of data representation, different kinds of graphs and various data filtering systems.

The first step of this work is related with the analysis of datasets structure. We currently focus our research on OD CSV files. *Piled Chart* has been created in order to show the structure of CSV files. It has currently some interesting features and we support the idea that he has potential to serve our purpose. However, it still also has several weaknesses and limitations. The main issue concerns the lack of the possibility to analyze and compare several files simultaneously. All new functionalities must take into account that the chart will have to be effective and intuitive. Hard job still has to be done. We are confident that we will provide a solution that will be able to show and compare effectively the structure of CSV files, as an important element of our complete visual OD integration solution.

References

1. Ziegler, P., Dittrich, K.R.: Three decades of data intecration - all problems solved? In: Jacquart, R. (ed.) Building the Information Society, pp. 3–12. Springer, US (2004)
2. Rivero, C.R., Schultz, A., Bizer, C., Ruiz, D.: Benchmarking the performance of linked data translation systems. In: LDOW (2012)
3. Vander Sande, M., Portier, M., Mannens, E., Van de Walle, R.: Challenges for open data usage: open derivatives and licensing. In: Workshop on Using Open Data (2012)
4. Houssos, N., Jörg, B., Matthews, B.: A multi-level metadata approach for a Public Sector Information data infrastructure. In: Proceedings of the 11th International Conference on Current Research Information Systems, pp. 19–31 (2012)
5. Otjacques, B., Cornil, M., Feltz, F.: Using ellimaps to visualize business data in a local administration. In: 2009 13th International Conference on Information Visualisation, pp. 235–240. IEEE (2009)

6. Rao, R., Card, S.K.: The table lens: merging graphical and symbolic representations in an interactive focus+ context visualization for tabular information. In: Proceedings of the SIGCHI Conference on Human Factors in Computing Systems, pp. 318–322. ACM (1994)
7. Davies, T.: Open data, democracy and public sector reform. A look at open government data use from data.gov.uk (2010)
8. Janssen, M., Charalabidis, Y., Zuiderwijk, A.: Benefits, adoption barriers and myths of open data and open government. Inf. Syst. Manag. **29**(4), 258–268 (2012)
9. McCusker, J.P., Lebo, T., Chang, C., McGuinness, D.L., da Silva, P.P.: Parallel identities for managing open government data. IEEE Intell. Syst. **27**(3), 55 (2012)
10. Bountouri, L., Papatheodorou, C., Soulikias, V., Stratis, M.: Metadata interoperability in public sector information. J. Inf. Sci. **35**, 204–231 (2008)
11. Edwards, P., Mayernik, M.S., Batcheller, A., Bowker, G., Borgman, C.: Science friction: data, metadata, and collaboration. Soc. Stud. Sci. 0306312711413314 (2011)
12. Gurstein, M.B.: Open data: empowering the empowered or effective data use for everyone? First Monday **16**(2) (2011)
13. Moreau, L., Groth, P., Miles, S., Vazquez-Salceda, J., Ibbotson, J., Jiang, S., Munroe, S., Rana, O., Schreiber, A., Tan, V., Varga, L.: The provenance of electronic data. Commun. ACM **51**(4), 52–58 (2008)
14. Rehm, G., Uszkoreit, H.: Multilingual Europe: A challenge for language tech. MultiLingual **22**(3), 51–52 (2011)
15. Martin, S., Foulonneau, M., Turki, S., Ihadjadene, M.: Risk analysis to overcome barriers to open data. Electron. J. e-Government **11**(2) (2013)
16. Bizer, C., Heath, T., Berners-Lee, T.: Linked data-the story so far. Int. J. Semant. Web Inf. Syst. **5**(3), 1–22 (2009)
17. Bauer, F., Kaltenböck, M.: Linked Open Data: The Essentials. Mono/monochrom edn., Vienna (2011)
18. Fox, P., Hendler, J.: Changing the equation on scientific data visualization. Science (Washington) **331**(6018), 705–708 (2011)
19. Duke, D.J., Brodlie, K.W., Duce, D.A., Herman, I.: Do you see what I mean? [Data visualization]. IEEE Comput. Graph. Appl. **25**(3), 6–9 (2005)
20. Reggi, L.: Benchmarking open data availability across Europe: the case of EU structural funds. Eur. J. ePractice **12**, 17–31 (2011)
21. Yu, H., Robinson, D.: The New Ambiguity of 'Open Government'. Princeton CITP/Yale ISP Working Paper (2012)
22. Harrison, T.M., Pardo, T.A., Cook, M.: Creating open government ecosystems: a research and development agenda. Future Internet **4**(4), 900–928 (2012)
23. Martin, M., Stadler, C., Frischmuth, P., Lehmann, J.: Increasing the financial transparency of european commission project funding. Semant. Web **5**(2), 157–164 (2014)
24. Zuiderwijk, A., Janssen, M.: Open data policies, their implementation and impact: a framework for comparison. Gov. Inf. Q. **31**(1), 17–29 (2014)
25. Shafranovich, Y.: Common format and MIME type for Comma-Separated Values (CSV) files (2005)
26. Maali, F., Cyganiak, R., Peristeras, V.: Enabling interoperability of government data catalogues. In: Wimmer, M.A., Chappelet, J.-L., Janssen, M., Scholl, H.J. (eds.) EGOV 2010. LNCS, vol. 6228, pp. 339–350. Springer, Heidelberg (2010)
27. Kalampokis, E., Hausenblas, M., Tarabanis, K.: Combining social and government open data for participatory decision-making. In: Tambouris, E., Macintosh, A., de Bruijn, H. (eds.) ePart 2011. LNCS, vol. 6847, pp. 36–47. Springer, Heidelberg (2011)

28. Charalabidis, Y., Lampathaki, F., Askounis, D.: Metadata sets for e-government resources: the extended e-government metadata schema (eGMS+). In: Wimmer, M.A., Scholl, H.J., Janssen, M., Traunmüller, R. (eds.) EGOV 2009. LNCS, vol. 5693, pp. 341–352. Springer, Heidelberg (2009)
29. Höchtl, J., Reichstädter, P.: Linked open data - a means for public sector information management. In: Andersen, K.N., Francesconi, E., Grönlund, Å., van Engers, T.M. (eds.) EGOVIS 2011. LNCS, vol. 6866, pp. 330–343. Springer, Heidelberg (2011)
30. Corlosquet, S., Delbru, R., Clark, T., Polleres, A., Decker, S.: Produce and consume linked data with Drupal!. In: Bernstein, A., Karger, D.R., Heath, T., Feigenbaum, L., Maynard, D., Motta, E., Thirunarayan, K. (eds.) ISWC 2009. LNCS, vol. 5823, pp. 763–778. Springer, Heidelberg (2009)
31. The EU Framework Programme for Research and Innovation. http://ec.europa.eu/programmes/horizon2020/en
32. ICT 2014 - Information and Communications Technologies. http://ec.europa.eu/research/participants/portal/desktop/en/opportunities/h2020/topics/87-ict-15-2014.html
33. DIRECTIVE 2003/98/EC OF THE EUROPEAN PARLIAMENT AND OF THE COUNCIL of 17 November 2003 on the re-use of public sector information. http://eur-lex.europa.eu/LexUriServ/LexUriServ.do?uri=OJ:L:2003:345:0090:0096:EN:PDF
34. Freedom of Information Act. http://foia.state.gov/
35. Government of South Australia State Records. http://www.archives.sa.gov.au/foi
36. Canada's Action Plan on Open Government. http://data.gc.ca/eng/canadas-action-plan-open-government
37. AGLS Metadata Standard. http://www.agls.gov.au/
38. Government Linked Data Working Group Charter. http://www.w3.org/2011/gld/charter
39. Linking Open Government Data. http://logd.tw.rpi.edu/
40. Creating Knowledge out of Interlinked Data. http://lod2.eu/Welcome.html
41. The open source data portal software. http://ckan.org/
42. Opening up Government. http://data.gov.uk/
43. Europe's Public Data. http://publicdata.eu/
44. Berlin Open Data. http://daten.berlin.de/
45. DKAN. https://drupal.org/project/dkan
46. data.gouv.fr, http://www.data.gouv.fr/
47. ToulouseMetropole.data [données publiques]. http://data.grandtoulouse.fr/
48. NANTES OUVERTURE DES DONNÉES - ouverture des données publiques. http://data.nantes.fr/accueil/
49. Strasbourg.eu et Communauté Urbaine. http://www.strasbourg.eu/fr
50. Arbres dans les parcs de la ville de Versailles. http://www.data.gouv.fr/en/dataset/arbres-dans-les-parcs-de-la-ville-de-versailles-idf

Security Issues in Distributed Data Acquisition and Management of Large Data Volumes

Alexander Kramer, Wilfried Jakob$^{(\boxtimes)}$, Heiko Maaß,
and Wolfgang Süß

Institute for Applied Computer Science, CN,
Karlsruhe Institute of Technology (KIT), P.O. Box 3640, 76021
Karlsruhe, Germany
`alexander.kramer@partner.kit.edu`,
`{wilfried.jakob,heiko.maass,wolfgang.suess}@kit.edu`

Abstract. The internet is faced with new application scenarios like smart homes, smart traffic control and guidance systems, smart power grids, or smart buildings. They all have in common that they require a high degree of robustness, reliability, scalability, safety, and security. This paper provides a list of criteria for these properties and focuses on the aspect of data exchange and management. It introduces a security concept for scalable and easy-to-use *Secure Generic Data Services*, called SeGDS, which covers application scenarios extending from embedded field devices for data acquisition to large-scale generic data applications and data management. Our concept is based largely on proven standard solutions and uses standard enterprise hardware. The first application deals with transport and management of mass data originating from high-resolution electrical data devices, which measure parameters of the electrical grid with a high sample rate.

Keywords: Data security · Data privacy · Scalable data exchange · Smart grid data management · Large data volumes · Data management

1 Introduction

New challenges arise from new smart application concepts demanding high rates of data exchange like smart traffic control and guidance systems, smart buildings, or smart power grids. As the latter shows a number of issues typical of such systems, we take a closer look at it. The old electrical supply system, which served mainly as a centralized power distribution network, is currently changing to a much more decentralized grid with a growing number of volatile energy sources. In addition, it is intended that the power consumption of more and more grid nodes can be influenced to some extent by a net supervisory system aiming at an increasing steadiness of the network load [1, 2]. Controlling the stability of such a power system is a much more complex task than the control of the old one and requires data acquisition in real time [3]. As a result, we have three types of data: Data on the consumption and feeding for billing purposes, data for consumption control, and data about the network status to control the stability of the network itself. All these data have in common that their confidentiality must be

M. Helfert et al. (Eds.): DATA 2014, CCIS 178, pp. 16–27, 2015.
DOI: 10.1007/978-3-319-25936-9_2

ensured. Data for billing and consumption control require privacy by nature and data about the network status must also be protected as they can be used for an attack on the network as well as for ensuring its stability [4]. Smart meters usually provide 1–15 min values consisting of cumulated power values over time. In contrast to that, data for network control are required in real time, which means at the level of a few seconds or less [3]. Both applications produce a large amount of data to be securely transferred, either because there is a large amount of data sources as in case of smart meters or because the update frequency is high.

Another important aspect is the dynamic nature of security and reliability. Both interact and the threats change over time. The more dissemination and diversity of any smart application increase, the larger does the vulnerability of the entire system grow. New threats will occur, which cannot be foreseen today. Thus, security measures are not a one-time business, but a permanent process throughout the entire life cycle of a network and of all of its components.

These considerations lead to the following requirements:

(a) Scalability:
New networks like smart power grids will start with a comparably small number of metering devices, but their number and data rates will grow over time.

(b) Heterogeneity:
Devices and software tools of different vendors used for different purposes and producing various data rates must be integrated.

(c) Suitability for different IT infrastructures

(d) High Reliability:
Online network control, for instance, requires an availability of close to 100 %.

(e) High Degree of Safety:
Many people will only accept smart grids as long as their privacy is secured. Data integrity must be ensured as well. The reliability of the power supply net is all the more essential the more a country is industrialized.

(f) Maintainability:
New security threats may require a fast reaction and, thus, it must be possible to quickly upload software updates to the affected components of the network. Furthermore, it must be possible to replace outdated security, transmission, or other methods and standards by up-to-date ones.

(g) Cost Effectiveness:
The smart power grid is to be a mass product. Acceptance of consumers requires low costs of the devices and services

(h) Restricted Access and Logging:
Access must be restricted to authorized personnel. Logging of all transactions is required to allow for a detection of attacks and misuse.

To handle diverse data and to facilitate different kinds of data processing, a flexible data management system is required. For this purpose, we developed our metadata-driven concept of Generic Data Services (GDS), see [5, 6], a first prototype of which was implemented for handling voltage measurement data of a very high resolution (12.8 kHz) needed for ongoing research projects [7, 8]. These devices are called

Electrical Data Recorders (EDR). Furthermore, the GDS stores the electric circuit plan of the Campus North of KIT, which is a classified document due to the shut down and operating nuclear installations, which have to be protected against terrorist attacks. The plan is required for the development of sub-models of the network, which serve as a basis for simulations and studies. Thus, GDS must provide a high degree of safety, especially as it is operated in an environment with a large number of users: More than 24,500 students and about 9,400 employees have access to the KIT LAN. This implies that administration of the comparably small number of GDS users must be separated completely from the user management of KIT.

In this paper we will introduce a concept for secure and reliable data transport, storage, and management, which will meet the above demands. It is based on standard hard- and software solutions and standardized interfaces, which considerably facilitates the fulfillment of a part of the listed requirements. In particular, the reliance on standardized interfaces follows directly from the heterogeneity and maintainability requirements. The rest of the paper is organized as follows. Section 2 gives a brief overview of related work. Our security concept is introduced in Sect. 3 and compared with the previously established requirements, while Sect. 4 reports about the first prototypic implementation. The last section summarizes the paper and gives an outlook on future work.

2   Related Work

IT security is a topic which is about as old as IT itself. Risks and threats grew with the growing capabilities of IT systems to today's cyber threats and challenges, see e.g. [9–12]. To secure data communication via the internet, several attempts have been made resulting in standards like IPSec [13, 14], TLS/SSL [15, 16], or the concept of virtual private networks (VPN) [13] based on these secure communication standards.

Berger and Iniewski [17] give an up-to-date overview of smart power grid applications and their technologies, including different communication techniques, and provide an in-depth discussion on the related security challenges. Mylnek et al. [18] propose a secure communication based on a selected encryption method, but it is intended to support only low-cost and low-power grid devices and thus, the concept lacks flexibility with respect to future requirements. Also IT infrastructure suppliers like Cisco [19], Juniper [20], or IBM in conjunction with Juniper [21] develop concepts for smart grid security and grid networking. A completely different approach is pursued in [22], where an incremental data aggregation method for a set of smart meters is proposed to protect user privacy. For further and permanently updated information, see the IEEE web portal on smart grids [23], where also security aspects are discussed.

A very good overview of the current state of the art of IT security is given in [24].

In Sect. 3.4 we use the concept of pseudonymization to protect the privacy of personal data and to allow e.g. processing for statistical purposes. Sweeney [25] proved that pseudonymization is insufficient in many cases. This is true, in particular, if other data of the persons in question are freely available, the combination of which with the pseudonymized data allows conclusions to be drawn with respect to the persons to be protected. He introduced the concept of k-anonymity, which is given if the information for each person contained in a data set cannot be distinguished from at least k-1 individuals whose information also appears in that set. The idea is continued in the GridPriv concept [26], which is also based on a set of data items per person, which might be used for re-identification and which are required by different service providers like energy suppliers, billing services, or network administrators. The more different data items are needed and used per person, the more possibilities exist for re-identification. Thus, it is an application dependent issue and we will come back to this later in Sect. 3.4.

3 SeGDS Concept

Before the security concept is described, we briefly introduce the GDS. It is an object- and service-oriented data management system designed to manage large amounts of data stored e.g. in the Large Scale Data Facility (LSDF) of KIT [27]. It is generic in so far, as it can deal with differently structured data and different kinds of storage systems. For this purpose, we defined three kinds of metadata: Structural metadata describe the structure of the data objects to be handled, while application metadata (AMD) are used to identify a data object. Thus, the AMD must be unique. It is left to the user to define which data shall serve for this identification purposes. It can be either a set of different user data or an identifier which is provided and managed by the application. The only requirement is its uniqueness. The third class of metadata is called organizational metadata (OMD) and it is used to manage the localization of data objects in storage systems and to handle security issues as described later in this section. Data objects are stored always as a whole and AMD are stored additionally as a metadata catalog. For the latter, the GDS uses its own data-base system, which is separated from the mass storage system used. A detailed description of the GDS in general and its metadata-based concept can be found in [6].

The concept of the Secure GDS (SeGDS) comprises:

- Secure data transport between clients and the GDS services, including authentication as described in Sects. 3.1 and 3.2.
- The aggregation of objects to be treated equally with respect to safety, see Sect. 3.3.
- Ciphering and pseudonymization discussed in Sect. 3.4.
- The management of users, user groups and access rights, see Sect. 3.5.

3.1 Overall Concept

The requirements a, b, d, f, and g from the above list suggest a solution based on standards rather than application-specific approaches. Cost effectiveness (g) of a scalable (a), heterogeneous (b), and highly reliable IT system, which can be updated quickly and adapted easily to new upcoming methods (f) requires the use of standards. To achieve a high level of safety (e), communication must be isolated and encrypted. At least in the beginning, the existing communication infrastructure has to be used to achieve low costs (g). Thus, we decided to use a virtual private network (VPN) based on standard hardware solutions to connect data acquisition devices like smart meters or more highly sophisticated devices like EDRs and user applications to the GDS via the present and insecure internet. This ensures scalability to a large extent, as the internet concept proved that it is highly expandable in the last 20 years. This also applies in the case of the establishment of a separate network from the internet, which may become necessary to avoid disturbances by load peaks of the public part of the network. As TLS/SSL has turned out to be mostly used for cyphering by clients, we recommend this secure communication method as well. The VPN shifts the burden of authentication from the application, here the GDS, to the VPN itself, as only registered users, who can authenticate themselves, are granted access (h). The practice shows that VPNs fit very well into different IT infrastructures and as they are independent of the structure of the data transferred, requirements *c* and *e* are also met. The growing amount of data (a) remains a critical point, especially since the data must be encrypted and decrypted. On the other hand, cyphering is a fundamental requirement regardless of the use of a VPN. As with the internet before, growing data volumes will require faster and/or more parallel hardware and communication lines.

Of course, the proposed solution has its price. But there must be a compromise between costs, on the one hand, and security and high reliability, on the other hand. Both cannot be achieved for free.

Figure 1 shows the overall concept. The clients are connected to the *VPN router* farm via the internet. The VPN routers share the traffic (load balancing) and pass it on to the alternatively usable *GDS Servers* and operate in failover mode, so that the service of a defective device can be taken over by others with the performance being reduced to some extent only. Authorization is done here by a *TACACS$^+$-Server* [28], which reads the user information, consisting among others of the user names and encrypted passwords, from an XML configuration file. The file is generated by the *GDS-Admin*

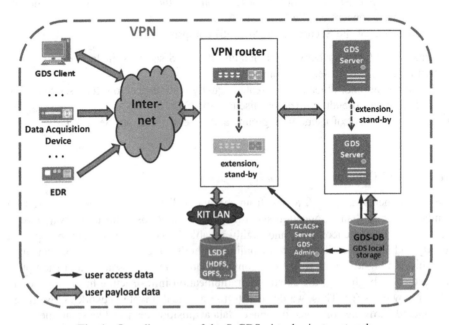

Fig. 1. Overall concept of the SeGDS virtual private network.

component after a change of the user list in the GDS data-base (*GDS-DB*). This results in a complete separation of the user management of VPN and GDS from the domain in which the SeGDS equipment is running. And it ensures that both components, the VPN and the GDS, work with the same user list. After successful authentication, different users are given different possibilities of access to the services of GDS according to the specifications of the access control lists. Data acquisition devices, for instance, will have access to appropriate services only, while human users or their applications may be granted extended or full access.

The *GDS-DB* shown in Fig. 1 is also used to store the already mentioned AMD and OMD of the data objects. The latter will be discussed in more detail in Sect. 3.5.

3.2   Secure Data Transport and Storage

The security of the data transported between the clients and the GDS is ensured by the encryption methods used by the VPN. The GDS decides according to given rules [6] where the data objects are stored. At present, either one of the file systems of the LSDF like the operating HDFS or the planned GPFS is used or the data are stored by the *GDS local storage* system. The latter also serves for experimental setups such as performance measurements, comparisons of different cyphers, or the like. According to the concept, the LSDF storage systems should be accessed via the VPN to ensure a maximum of safety. But this must be left to a future enhancement, as will be described in Sect. 4.

Stored data must be protected against loss and change. The first threat is covered by the standard backup procedures of the computer center hosting the LSDF or the local storage of the GDS. Alterations of data can be detected by cryptographic hash values resulting from algorithms like SHA-2 or the upcoming SHA-3 [29], which are computed and saved when the data are stored. When reading the data, its integrity is checked by calculating the hash value again and comparing it with the stored one. In case of corrupted data, the standard data backups of the data center, in our case the LSDF, can be used to restore the original version.

3.3   Data Objects and Object Sets

We assume that many elementary *data objects* can be treated equally in terms of access rights and encryption. These objects form an *object set*. For the sake of generality, object sets may also have only one or a few objects, but we do not expect this to be the ordinary case. Every elementary data object belongs to exactly one object set.

3.4   Pseudonymization and Ciphering

In many cases, a pseudonymization of personal data may be considered a sufficient measure to provide privacy and to allow e.g. processing for statistical purposes. It is assumed, of course, that the pseudonymized data cannot be reconstructed, which is an application-dependent question.

The application at hand, which is described in greater detail in Sect. 4, creates data objects, which consist of a time series of the measured data, date and time of the recording, the GPS coordinates of the metering device, and the exact position within the power grid. From the latter, an allocation to homes, apartments, or offices, and, thus, to people can be made. Consequently, the data must be treated and protected as personal data. To avoid this, the position is substituted by a pseudonym and the GPS coordinates will be omitted in the future. The question of safety of these pseudonyms is whether the remaining data are suitable for the reconstruction of the measurement site, see also [25]. For this, the power supply would have to be manipulated locally at a certain time, such that it generates a uniquely assignable measurement result, by which the recording can be identified. Due to the nature of the power supply system, such interferences would always affect many or at least several consumption points. Hence, this is not a practical method to compromise the pseudonyms for this application. And more data are not available.

If, in other cases, pseudonymization is not sufficient to protect privacy and/or if it is required by the user, all data objects of a set may be stored encrypted to provide security against unauthorized and illegal access to an external mass storage system like the present HDFS of the LSDF. There is a key per set, which is administrated by the GDS. The GDS performs encryption and decryption, so that the ciphering is completely transparent to the user except that access may slow down.

An additional security level can be provided, if the user application does the ciphering and the data objects arrive at the GDS already encrypted. In this case, the GDS needs the identifying metadata in clear text only.

Anonymization is another issue that will be dealt with. Since the current applications do not allow anonymity, but only pseudonyms, anonymization is processed later.

3.5   Users, Groups and Access Rights

As with many other data administration systems, we have *users*, who may be merged into *groups*, provided that they have the same *access rights* to object sets.

Users and their Properties. Every registered application or person is a *user*, who may be a member of one or more groups. It is distinguished between ordinary users and *administrators*, who have special rights, as will be explained later.

Each object set is owned by exactly one user. Users may, but need not possess one or more object sets.

Every user has a default object set, to which new data objects belong, provided that the writing GDS service is not told to use a different one. The default object set may, but needs not be possessed by the user it is associated with. This means that it is possible that a user stores data objects belonging to an object set, which is not his own. The idea behind this is that it may be meaningful for some automatic data sources to store their objects into the same set, which belongs e.g. to the operator controlling these sources. For reasons of security, every device acts as a separate so called device-user, which can log-in at the same time only once. Thus, a further attempt to login can be detected easily. This does not limit the scalability, as new device users can be cloned quickly from a predefined standard schema.

Users may be permanent or temporary. This is also motivated by the automated data sources like the EDRs or other data acquisition devices, which may send data for a limited duration only. This possibility of time-limited validity of users may also be used to grant access to persons for a limited period of time, for example to students doing an internship. As users may possess object sets and object sets must be owned by someone, a user may not be deleted automatically upon deactivation. Thus, the system must not only distinguish between permanent and temporary users, but also among temporary users who are active, passive and waiting for their activation, or passive due to time-out. Temporary, expired users remain in the system until they are erased by an administrator as described below in the section on management.

User Groups. A *group* consists of users with the same access rights to some object sets in each case. A group consists of one user at the minimum and has access to at least one object set. Object sets can be accessed by no, one, or more groups. As an object set must always be possessed by a user, there is still access to a set, even in the case of no group being left with permissions to access it.

Access Rights. There are three basic access rights:

- Read Permission
  In addition to reading all data objects of an object set, the creation of lists according to different criteria (search lists) is allowed.
- Write Permission
  Allows creating a new data object.
- Delete Permission
  Permission to delete single data objects or an entire object set, including its data objects.

For updates of already existing objects, both rights the read and the delete permissions are needed. These three access rights determine the access capabilities of a user regarding his own data sets or of a group concerning any data sets. Regarding his own data sets, a user can change the access rights of himself as the owner or of a group.

In addition to these user-changeable access rights to data sets, every user has a set of so-called *static rights*, which can be controlled by administrators only. They consist of the same access rights as before and can generally switch on or off a particular access right of a user. The rationale for that is to have a simple possibility for administrators to reliably limit the rights of a user without the need to consider his group rights and without allowing him to modify that even in case of his own object sets.

Management of Users, Groups, and Object Sets. Administrators are users with special additional capabilities. Only administrators can manage users and groups. They can give themselves all access rights to object sets and they can change the ownership of object sets as well as the access rights of the new owner. This ensures maintainability of the GDS even in case of permanent absence of a user: All the data sets of such a user can be modified so that the data remain usable. For reasons of security, there is one thing administrators cannot do as with other systems: They cannot retrieve the password of a user in plaintext. But, of course, they can reset it.

The exclusively administrator controlled functions are managed by a local tool within the VPN, as is indicated by *GDS-Admin* in Fig. 1. It offers the following functions to administrators

- Creation of a user and assignment of the initial object set. If this is a new set, it must be created also to complete the creation of that user. For temporary users, the given start and end times are checked for plausibility: The start time must not be in the past and must be earlier than the end time.
- Alteration of user data.
- Deletion of a user. This requires that he does not possess any object sets. It implies removal from all groups the user was a member of.
- Creation and deletion of a group.
- Addition of a user to a group.
- Deletion of a user from a group.

The following further functions are available to administrators locally and remotely as a service for common users. If used by an administrator they can be applied to any user, but an ordinary user can perform them only on own data objects, objects sets, memberships, or user data. As this restriction is valid for all functions below, it is not repeated for reasons of linguistic simplicity:

- Granting, deleting, or changing access rights to an object set for a group.
- Changing of access rights of an owner to his object sets.
- Creation and deletion of an object set. Only empty object sets are erasable. For a newly created object set owner access rights must be given.
- Transfer of the ownership of an object set to another user.
- Transfer of data objects to another object set. If applied by an ordinary user, he must be the owner of the source object set.
- Listing functions for users and groups and their access rights.
- Change of a password.

User and Rights Administration. As pointed out above, the management of the VPN and GDS users is completely separated from the user management of the IT infrastructure which hosts both VPN and GDS. The list of VPN users, the TACACS$^+$-server relies on is generated by the user administration tool of the GDS. Therefore, the services of GDS can be used only by users, who have authenticated themselves before access was granted. Furthermore, the administration tool itself can be accessed locally only. We think that the overall security is further enhanced by these measures.

4 Current Implementation

Figure 2 shows the current prototypic implementation, which at present is mainly used to manage data objects generated by the EDRs. In the future, also data of the Electrical Grid Analysis Simulation Modeling and Visualization Tool (*e*ASiMoV), see [5, 7] will be managed. The VPN is realized using a Cisco router, mainly because we have an

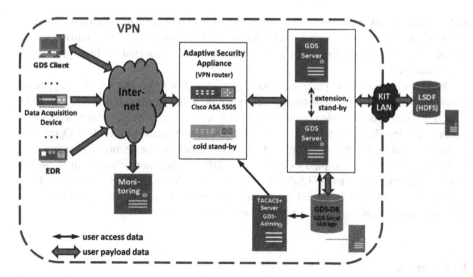

Fig. 2. First implementation of the SeGDS concept based on one active Cisco ASA 5505 and six parallel GDS servers. A second ASA is available as cold stand-by to replace the active one in case of a breakdown. For cost reasons, the LSDF is connected through the KIT LAN.

existing infrastructure based on Cisco hardware and the respective licences and everything else would be more expensive. Nevertheless, other manufacturers like Juniper or Checkpoint can be used alternatively, of course. At present, we use one Cisco ASA 5505 with a back-up device of the same type in cold stand-by. Unfortunately, this is a bottleneck due to a limited budget for the prototype.

The main structural difference to the concept shown in Fig. 1 is that the HDFS file server is accessed via KIT LAN outside of the VPN, which is done mainly for cost reasons. This solution is justifiable as long as the stored data are pseudonymized, as it is the case with the EDR data. The planned integration of a GPFS file server will be done more securely via ssh or scp and/or within the VPN.

There is a special client called *Monitoring*, which was added to the current implementation. It is based on RDP (Remote Desktop Protocol) and serves as a tool for supervising the EDRs. A list of connected EDR devices, including performance information about the acquisition hardware and data transfer, is created. If necessary, EDRs can be restarted. Since the monitoring is only used within the VPN, the known security weaknesses of RDP can be accepted at this stage of application. The monitoring tool helps to detect malfunctions of the EDRs and to fix them by restarting also from outside of the KIT campus.

5 Summary and Outlook

We have given a list of criteria for a secure, reliable, scalable, and generic data exchange and management system and demonstrated how they can be met by standard solutions. The preference of standard solutions results in both comparably low prices and synergy effects with other applications in terms of technical development and new standards and the discovery of vulnerabilities and their elimination. We introduced an overall concept of secure generic data services and reported a first prototypic implementation.

Meanwhile, we have gathered first experience from the use of the suggested hardware-based VPN. For a laboratory setup like ours, the solution based on ASA 5505 does not seem to be flexible enough and it requires too much maintenance effort for a research group. Thus, we will compare it with an implementation based on OpenVPN in the near future. Further development will concentrate on the secure integration of a GPFS file server. We also plan to enlarge the VPN so that more clients can be added and the communication to the LSDF is integrated. Parallel to that, the robustness of the security measures will be tested by supervised intrusion attacks. The quality of the approach will be investigated in various performance-tests.

References

1. Fed. Min. of Economy and Energy: E-Energy: Startpage. http://www.e-energy.de/en/. Last accessed on May 14, 2014
2. U.S. Dept. of Energy: Home| SmartGrid.gov. https://www.smartgrid.gov/. Last accessed on May 14, 2014
3. Bakken, D., Bose, A., Hauser, C., Whitehead, D., Zweigle, G.: Smart Generation and Transmission with Coherent, Real-Time Data. Proc. IEEE **99**, 928–951 (2011)
4. Smart Grid Security. Annex II: Security aspects of the smart grid (2012)
5. Maaß, H., Çakmak, H.K., Süß, W., Quinte, A., Jakob, W., Stucky, K.-U., Kuehnapfel, U.G.: Introducing the electrical data recorder as a new capturing device for power grid analysis. Applied Measurements for Power Systems (AMPS). 2012 IEEE International Workshop on, pp. 1–6. IEEE, Piscataway (2012)
6. Stucky, K.-U., Süß, W., Çakmak, H.K., Jakob, W., Maaß, H.: Generic Data Management Services for Large Scale Data Applications. to be published (2015)
7. Maaß, H., Çakmak, H.K., Bach, F., Kuehnapfel, U.: Preparing the electrical data recorder for comparative power network measurements. accepted for publication. In: 2014 IEEE International Energy Conference and Exhibition (ENERGYCON). IEEE, Piscataway (2014)
8. Bach, F., Çakmak, H.K., Maaß, H., Kuehnapfel, U.: Power grid time series data analysis with Pig on a hadoop cluster compared to multi core systems. In: Stotzka, R., Milligan, P., Kilpatrick, P. (eds.) Proceedings of the 2013 21st Euromicro International Conference on Parallel, Distributed, and Network-based Processing, pp. 208–212. IEEE, Piscataway (2012)
9. Menezes, A.J., van Oorschot, P.C., Vanstone, S.A.: Handbook of applied cryptography. CRC Press, Boca Raton (1997)
10. Ferguson, N., Schneier, B., Kohno, T.: Cryptography Engineering, Design principles and practical applicationns. Wiley, Indianapolis (2010)
11. Partida, A., Andina, D.: IT security management. Springer, Dordrecht, London (2010)
12. Yu, T., Jajodia, S. (eds.): Secure Data Management in Decentralized Systems. Springer, New York (2007)
13. Doraswamy, N., Harkins, D.: IPSec. The new security standard for the Internet, intranets, and virtual private networks. Prentice Hall PTR, Upper Saddle River (2003)
14. Stallings, W.: Network Security Essentials: Applications and Standards. Prentice Hall, Upper Saddle River (2013)

15. Rescorla, E.: SSL and TLS: Building and Designing Secure Systems. Addison-Wesley, Harlow (2000)
16. Oppliger, R.: SSL and TLS: Theory and Practice. Artech House, Boston (2009)
17. Berger, L.T., Iniewski, K.: Smart Grid Applications, Communications, and Security. Wiley, Hoboken (2012)
18. Mlynek, P., Misurec, J., Koutny, M., Raso, O.: Design of secure communication in network with limited resources. In: 2013 4th IEEE/PES Innovative Smart Grid Technologies Europe (ISGT EUROPE), pp. 1–5. IEEE, Piscataway (2013)
19. Cisco Systems: Grid Security - Industry Solutions. http://www.cisco.com/web/strategy/energy/smart_grid_security.html. Last update on January 1, 2011; last accessed on May 13, 2014
20. Juniper Networks: Energy and Utilities - Smart Grid Security Solution. http://www.juniper.net/as/en/solutions/enterprise/energy-utilities/. Last accessed on May 13, 2014
21. IBM Corporation: IBM and Alliance - Energy and utilities solutions from IBM and Juniper Networks - United States. http://www.ibm.com/solutions/alliance/us/en/index/juniper_energy.html. Last update on May 8, 2014; last accessed on May 13, 2014
22. Li, F., Luo, B., Liu, P.: Secure and privacy-preserving information aggregation for smart grids. IJSN **6**, 28–39 (2011)
23. IEEE: Smart Grid Experts, Information, News & Conferences. http://smartgrid.ieee.org/. Last accessed on May 16, 2014
24. Eckert, C.: IT-Sicherheit: Konzepte Verfahren, Protokolle. Oldenbourg, München (2012)
25. Sweeney, L.: k-anonymity: a model for protecting privacy. Int. J. Unc. Fuzz. Knowl. Based Syst. **10**, 557–570 (2002)
26. Stegelmann, M., Kesdogan, D.: GridPriv: A Smart Metering Architecture Offering k-Anonymity. In: IEEEE (ed.) 2012 IEEE 11th International Conference on Trust, Security and Privacy in Computing and Communications Communications (TrustCom), pp. 419–426. IEEE, Piscataway (2012)
27. García, A., Bourov, S., Hammad, A., van Wezel, J., Neumair, B., Streit, A., Hartmann, V., Jejkal, T., Neuberger, P., Stotzka, R.: The large scale data facility: data intensive computing for scientific experiments. 25th IEEE International Symposium on Parallel and Distributed Processing, IPDPS 2011. Workshop Proceedings, pp. 1467–1474. IEEE, Piscataway, NJ (2011)
28. Finseth, C.: An Access Control Protocol, Sometimes Called TACACS. https://tools.ietf.org/html/rfc1492. Created: July 1993; last accessed on October 22, 2014
29. NIST: SHA-3 Standardization. http://csrc.nist.gov/groups/ST/hash/sha-3/sha-3_standardization.html. Last update on April 7, 2014; last accessed on May 12, 2014

Large-Scale Residential Energy Maps: Estimation, Validation and Visualization Project SUNSHINE - Smart Urban Services for Higher Energy Efficiency

Umberto di Staso[1]([✉]), Luca Giovannini[2], Marco Berti[1],
Federico Prandi[1], Piergiorgio Cipriano[2], and Raffaele De Amicis[1]

[1] Fondazione Graphitech, Via alla Cascata 56/C, Trento, TN, Italy
{umberto.di.staso,marco.berti,federico.prandi,
raffaele.de.amicis}@graphitech.it
[2] Sinergis Srl, Via del Lavoro 71, Casalecchio di Reno, BO, Italy
{luca.giovannini,piergiorgio.cipriano}@sinergis.it

Abstract. This paper illustrates the preliminary results of a research project focused on the development of a Web 2.0 system designed to compute and visualize large-scale building energy performance maps, using: emerging platform-independent technologies such as WebGL for data presentation, an extended version of the EU-Founded project TABULA/EPISCOPE for automatic calculation of building energy parameters and CityGML OGC standard as data container. The proposed architecture will allow citizens, public administrations and government agencies to perform city-wide analyses on the energy performance of building stocks.

Keywords: CityGML · WebGL · 3D city model · Smart city · Interoperable services · Energy efficiency · City data management · Open data · Building energy efficiency · Energy performance index

1 Introduction

One of the current hottest topics in the information technology research area is certainly the one about "smart-cities". But, what is a smart-city? Many definitions exist in the current literature [3, 6–8, 18] and all of them have a factor in common: the existence of an underlying ICT infrastructure that connects the physical infrastructure of the city with web 2.0 capabilities and enables innovative solutions for city management, in order to improve sustainability and the quality of life for citizens.

The energy consumption efficiency of residential houses is an important factor having an impact on the overall city ecosystem and quality of living and it would greatly benefit from an ICT-enabled smart approach. In fact, increasing building energy efficiency would not only mean a cut-down in energy expense for citizens, but would also have an impact on the overall production of $CO2$ at energy plants and also, even if less intuitively, on the city air pollution. Among of the major causes of poor air quality

© Springer International Publishing Switzerland 2015
M. Helfert et al. (Eds.): DATA 2014, CCIS 178, pp. 28–44, 2015.
DOI: 10.1007/978-3-319-25936-9_3

are actually industries and domestic heating systems, not the quantity of vehicles circulating in the urban area, as one might more typically think [6].

So, what kind of smart service can be designed in order to support the increase of building energy efficiency and improve the city quality of life in this respect? What we present in this paper is a specific answer to this question. The paper will illustrate the concept and the development of smart services, which allow the assessment of the energy performance of all the residential buildings in a city, its validation and visualization in a format accessible to citizens and urban planning experts alike.

The development of these services is part of the scope of the SUNSHINE project (Smart UrbaN ServIces for Higher eNergy Efficiency, www.sunshineproject.eu), that aims at delivering innovative digital services, interoperable with existing geographic web-service infrastructures, supporting improved energy efficiency at the urban and building level. SUNSHINE smart services will be delivered from both a web-based client and dedicated apps for smartphones and tablets. The project is structured into three main scenarios:

- **Building Energy Performance Assessment:** Automatic large-scale assessment of building energy behavior based on data available from public repositories (e.g. cadaster, planning data etc.). The information on energy performances is used to create urban-scale maps to be used for planning activities and large-scale energy pre-certification purposes.

- **Building Energy Consumption Optimization:** Having access via interoperable web-services to real-time consumption data measured by smart-meters and to localized weather forecasts, it is possible to optimize the energy consumption of heating systems via automatic alerts that will be sent by the SUNSHINE app to the final users.

- **Public Lighting Energy Management:** Interoperable control of public illumination systems based on remote access to lighting network facilities via interoperable standards enables an optimized management of energy consumption from a web-based client as well as via the SUNSHINE app.

This paper focuses on the preliminary results for the first of the three scenarios. The aim of the service for Building Energy Performance Assessment is to deliver an automatic large-scale assessment of building energy behavior and to visualize the assessed information in a clear and intuitive way, via the use of what we call energy maps.

Energy maps will be made publicly available via a 3D virtual globe interface based on WebGL [14] that leverages on interoperable OGC standards, allowing citizens, public administrations and government agencies to evaluate and perform analysis on the building energy performance data. The presentation of these energy performance data in a spatial-geographic framework provides a global perspective on the overall performance conditions of the residential building stock as well as on its fluctuations on the neighborhood and block scale. It is thus the key for an efficient maintenance planning to increase the overall energy efficiency, allow citizen to save more money and, ultimately, improve the quality of living of the city.

2 Energy Maps: State of the Art

The current availability of relevant technologies and standards has encouraged the development of many research projects in the area of building energy performance estimation based on publicly available data with the aim of creating energy map.

The main challenge in this task is related to effectively providing data for the whole city area. For example, building certificates, adopted by many of EU countries to describe building efficiency, can provide a very detailed insight on building energy properties, but on the other hand, these certifications are not mandatory for all the residential buildings and their availability is thus very sparse.

So, given the fact that publicly available data generally do not include all the information needed for the energy performance calculation, one of the most common approaches to energy map creation is to estimate the missing information in a reliable way, using the basic input data that is typically available, such as building geometry, building use, construction year, climatic zone, etc. A solid example of this approach is described in [15], where the City Geography Markup Language (CityGML) standard [9] is used to semantically describe the set of objects that compose the urban environment, a building typology database is exploited to statistically estimate the energy performance properties of buildings and, finally, an Application Domain Extensions (ADE) to the CityGML model is defined to store the estimated information for each building [4, 11, 12].

A radically different approach is described in [10], where thermal images acquired by airborne thermal cameras are used to measure the heating lost by buildings via their roofs and windows and from that the energy performance of the buildings is estimated.

Both approaches have merits and deficiencies. In the former case, input data are publicly available, requiring no additional cost; however, having to rely on typological databases to estimate the most of the energy parameters yields a result that is typically not very statistically reliable at the building scale and is usually confined to residential buildings (where performance typologies are easier to define). Moreover, the overall software architecture is typically desktop based, so the access to the results is often limited to a small number of users with advanced GIS skills. Another limit is related to the dissemination and exploitation activities of the computed results: for performance reasons, the visualization is commonly provided via a conversion to KML [19], where the link between the building performance data and its geometry is color-coded in each building-style parameter and the other information stored in the starting CityGML file is lost.

In the thermal image approach, instead, all the building use typologies are taken into consideration, but the cost to collect thermal images to cover an entire city is hardly negligible. Furthermore, only the roof surface is evaluated in terms of energy performance, ignoring the full contribution of walls. Moreover, the use of proprietary standards does not encourage the adoption of the same solution by the research community.

The approach we present in this article belongs to the typological kind, but makes an effort to reduce the common drawbacks that have been delineated. As will be described in more details in the following sections, our approach is in effect hybrid, leveraging on the outcomes of project TABULA-EPISCOPE [2] but limiting the use of building parameters estimated typologically.

3   SUNSHINE Approach

3.1   System Architecture

In this chapter, the system architecture of the SUNSHINE platform is presented. As reported in the introduction, the SUNSHINE project covers three different scenarios; however, given the focus of the paper on the first scenario, the system architecture description has been rearranged to focus on the components that are meaningful in this context.

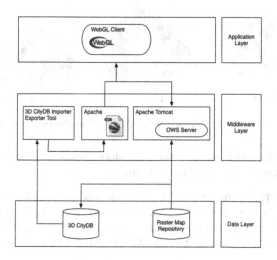

Fig. 1. System architecture.

The chosen approach for this scenario was that of leveraging on a building typology database and the system architecture that has been designed to comply with it (see Fig. 1) is based on a Services Oriented Architecture (SOA) with three tiers (data, middleware and application layers). A SOA-based approach allows accessing to the resources through a middleware layer in a distributed environment and thus avoids single access-points limitations that are instead typical for desktop-based architectures.

Data Layer. The bottom level of the SUNSHINE system architecture is aimed at storing geometry and semantic information about buildings and thematic raster data. The two fundamental components of this layer are the 3D CityDB [5] and the raster map repository.

The 3D City Database [17] is a free 3D geo database designed to store, represent, and manage virtual 3D city models on top of a standard spatial relational database, up to five different Levels of Detail (LoDs, see Fig. 2). The database model contains semantically rich, hierarchically structured, multi-scale urban objects facilitating GIS modeling and analysis tasks on top of visualization. The schema of the 3D City Database is based on the CityGML [9] standard for representing and exchanging virtual 3D city models.

The 3D City Database is described via a PostGIS relational database schema and specific SQL scripts are provided to create and drop instances of the database on top of a PostGIS DBMS.

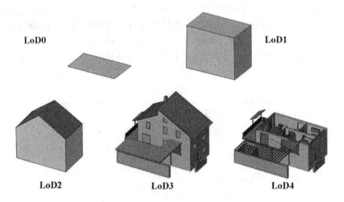

Fig. 2. CityGML LOD example [courtesy of http://www.simstadt.eu].

The raster map repository is a data file store aimed at containing geo-located orthophoto and elevation files in raster file format.

Middleware Layer. The middleware layer is the core component of the SUNSHINE platform. Its duty is to manage the connection between the application layer and the data layer, providing access to the resources stored in databases and map repository. The middleware layer is composed by the 3D CityDB [5] Importer/Exporter Tool and Apache Tomcat.

The 3D CityDB Importer/Exporter Tool allows interacting with the 3D City Database or external data. The Apache Tomcat [1] is an open source software implementation of the Java Servlet and JavaServer Pages technologies.

Application Layer. A further challenge that the SUNSHINE project took into high consideration is the dissemination and exploitation of the reached results. A smart-city will become smarter only if all the involved stakeholders (citizens, public administrations and government agencies) are aware about the outcomes of the research activities in that particular scope. For this reason, a great effort was put in designing and implementing a client platform that would be usable by the majority of devices, both mobile and desktop-based.

To achieve the widest dissemination possible for the project's results, the emerging WebGL technology [14] has been employed, in conjunction with HTML5 elements, as the main component of the application layer. WebGL is a cross-platform royalty-free web standard for a low-level 3D graphics API based on OpenGL ES 2.0, exposed through the HTML5 Canvas element as Document Object Model interface.

3.2   Energy Map Generation

The aim of this section is to describe how each building is associated with an energy class index. The energy class is an index describing the energy performance of a building and it is usually computed from a series of detailed information on building energy properties that are not available in general as public domain data. Publicly available data is usually limited to more basic information, such as building geometry, year of construction, number of building sub-units, etc. So, the approach we followed was to estimate the energy parameters needed for performance calculation from the few publicly available data.

More specifically, the data necessary to the estimation are:

- **Geometrical Data:** i.e. footprint, height, number of floors, etc. From these data, using specific geoprocessing procedures, other geometrical properties are derived, such as the building volume of the extent of the building walls shared with neighboring buildings.
- **Thermo-physical Data:** i.e. period of construction, prevalent building use, refurbishment level. From these data, using a typological approach and leveraging on a sample of representative buildings for the different thermo-physical typologies, the thermal properties of each building are estimated, such as U-values of envelope elements and the percentage of windowed surface.
- **Climatic Data:** i.e. the extent of the heating season and the average external temperatures. These data are derived from national and local directives.

As the following sub-sections will describe with more details, using the geometrical, thermo-physical and climatic data and applying a simplified computation procedure based on ISO 13790 and ISO 15316 (international standard protocols for the energy sector), the following parameters are computed for each residential building:

- The energy need for heating;
- The energy need for heating and domestic hot water;
- The corresponding index for energy performance.

There are some relevant aspects to highlight about this approach. The first is related to the fact that the building typological classification currently applies to residential buildings only and thus cannot be used to assess the energy performance of buildings with a predominant use that is other than residential (commercial, administrative, industrial, educational, etc.). As a consequence, the energy map itself will carry information only for residential buildings. This seems to us a reasonable compromise as residential buildings are among the major causes for energy consumption and air pollution [6].

A second important aspect is the use of thermo-physical typologies in order to estimate building properties that would be otherwise hardly obtainable on a large scale without employing a great deal of resources (money and time) and whose knowledge is instead necessary to determine an estimate of energy performance. The definition of these typologies is based on the results of project TABULA [13], integrated and extended to adapt to the specificities of SUNSHINE. Project TABULA defined a set of building typologies for each of the countries participating into the project, basing on 4 parameters: country, climate zone, building construction year, building size type (i.e. single family house, terraced house, apartment block, etc.). A building stereotype, described in all its geometrical and thermo-physical properties, is associated to each class, with the aim of representing the average energy behavior for buildings of that class. So, if the 4 parameters are known, than it is possible to associate the building to a specific typology class and thus to its estimated energy performance class.

As anticipated, the energy performance estimation approach developed in SUN-SHINE is hybrid, it differs from the typological approach followed in TABULA because it limits the use of typological estimation only to the thermo-physical properties, using instead a deterministic approach for the geometrical properties, measured or computed, of the involved buildings. A fully typological approach has in fact the intrinsic limitation that the statistical significance of the performance estimation directly proportional to the scale at which the approach is applied, so very low at the scale of the single building. A hybrid approach that takes into account the real geometrical properties of the building makes the estimate of the building energy performance more accurate. The validation step that will be described in Sect. 3.3 makes this approach even more robust.

Input Data Model. To implement the conceptual idea explained in the previous section, a set of data related to the buildings in the pilot urban scenarios of the cities of Trento and Cles, Italy, has been collected and the SUNSHINE energy map estimation workflow has been executed on it. Table 1 contains the data model created to collect such buildings data.

Table 1. Energy maps input data model.

Attribute name	Type
Building identifier	string
Building geometry	geometry
Begin construction year	integer
End construction year	integer
Building height	real
Floors	integer
Average floor height	real
Refurbishment level	{no/standard/adv}
Use	string

Energy Performance Estimation. This section is the core of this article, here is described the automatic process for large-scale building energy performance estimation, that is the base for the energy maps. The workflow illustrated in this section is implemented by using the following software:

- A Postgres relational database with PostGis extension, where an input shapefile, structured in accordance with the input data model, is loaded by the SHP2PG tool.
- FME 2014 Professional edition, ETL tool for manipulating spatial data. The same results can be achieved also by the use of free ETL tools, for example GeoKettle.

The procedure implemented by FME 2014, connected with the PostGIS database where input shapefiles are loaded, is represented in Fig. 3.

More in details:

1. The workflow starts initializing the system;
2. For each building, all the data gathered in the input data model is used and, in addition, additional geometries parameters are computed such as area, perimeter, shared and exposed walls perimeter;
3. The building typology, according to the categories provided by TABULA, is estimated with the algorithm described in Fig. 4;
4. Using the previous estimated parameters, it is possible to query the TABULA database in order to obtain the set of U_VALUES in accordance with the climatic zone, typology, construction year and refurbishment level;
5. Having the set of U_VALUES and the real geometry proprieties it will be possible to estimate the Energy Performance Index (EPI) according with the EN ISO 13790 regulation.
6. An output shapefile extending the input data model with the new geometrical and thermo-physical data is produced. More details regarding the output data model will be provided in the next section.

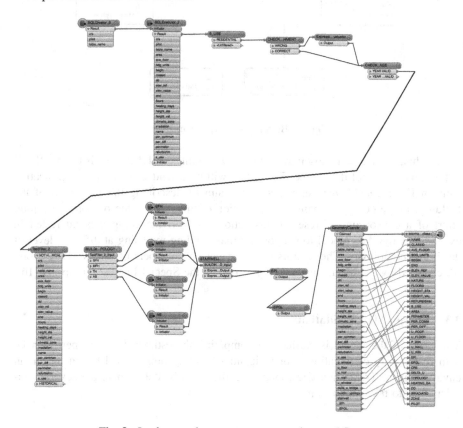

Fig. 3. Implemented energy maps generation workflow.

Output Data Model. The workflow illustrated in the previous section produces a new data model that is described in Table 2. As it is possible to see, the workflow's output is an extension of the input data model, where thermo-physical and additional geometrical data are provided. It is important to underline the fact that the set of U_VALUES are the ones provided by TABULA, while other data are computed starting from geometrical and topological information.

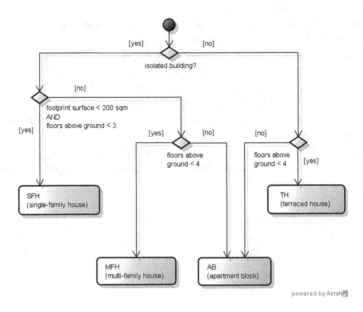

Fig. 4. Building size type estimation.

The final data model chosen to host the 3D buildings and their data is CityGML [9] at the LoD-1 level of detail (see Fig. 2) that will be extended with a new Application Domain Extension (ADE) defined with the aim of describing the properties of the buildings in the energy domain. The definition of this ADE is the outcome of a joint effort of a consortium of research institutions and private companies interested in expanding CityGML model to support thermal building modeling at different level of detail and complexity. The authors of this paper are members of this working group which is led by the Modeling Working Group of the Special Interest Group 3D [16].

3.3   Energy Map Validation

Validation of the model is carried out comparing the estimated energy performance with real energy certifications, for residential building built after 1900 in the urban environment of Trento and Cles, located in the north of Italy. There are several considerations to take into account:

Table 2. Energy maps output data model.

Attribute name	Type
Building identifier	string
Building geometry	geometry
Begin construction year	integer
End construction year	integer
Building height	real
Floors	integer
Average floor height	real
Refurbishment level	{no/standard/adv}
Use	string
Area	real
Perimeter	real
Shared perimeter wall	real
Exposed perimeter wall	real
U_roof	real
U_floor	real
P_win	real
U_wall	real
U_win	real
EPI	real
EPGL	real
CRS	string
Delta_U_bridge	real
Building_typology	int
Heating_days	int
Pilot_id	int
Irradiation	real
Climatic_zone_id	int

1. The SUNSHINE Workflow provides an estimation based on the whole building geometry while energy certifications are apartment-based;
2. The apartment position (ground floor/middle floor/last floor) influences the heat loss. For this reason, the energy performance estimation workflow is refined as follows and performed for the three above mentioned conditions. According by EN ISO 13790:

$$Q_{H,nd} = 0,024 \cdot (Q_{H,tr} + Q_{H,ve}) \cdot t - \eta_{H,gn} \cdot (Q_{int} + Q_{sol}) \tag{1}$$

$$Q_{H,tr} = H_{tr,adj} \cdot (\theta_i - \theta_e) \tag{2}$$

$$H_{tr,\text{adj}} = \sum \left(\alpha_i \cdot A_{env,i} \cdot U_{env,i} \cdot b_{tr,i} \right) + \Delta U_{tb} \cdot \sum \left(\alpha_i \cdot A_{env,i} \right) \qquad (3)$$

With:

$$\alpha_i = 1 \text{ for walls and windows;}$$
$$\alpha_i = \{0, 1\} \text{ for roof or floor;}$$

In particular:

$\alpha_{whole\ building,i} = 1$ for all elements;

$\alpha_{ground\ floor,i} = 1$ for {walls, windows, floor}; $\alpha_{ground\ floor,i} = 0$ for roof;

$\alpha_{middle\ floor,i} = 1$ for {walls, windows}; $\alpha_{middle\ floor,i} = 0$ for {floor, roof};

$\alpha_{last\ floor,i} = 1$ for {walls, windows, roof}; $\alpha_{last\ floor,i} = 0$ for floor;

1. Different software used to calculate the energy performance index produces similar, but not identical, results. The differences between these results are variable and can arrive to 20 % in the worst case.

Table 3 reports a subset of the whole set of buildings involved in the two urban environments where, for each entry, the following information is reported:

- Building typology, in according with the TABULA classification;
- City (TN/CL);
- Number of floors;
- Construction year;
- Real EPI [KWh/(m^2 year)]: energy performance index provided by real certificates;
- Estimated EPI – whole building [KWh/(m^2 year)]: estimated by the analysis on the entire building. This is the output of the unrefined SUNSHINE energy map workflow;
- Final EPI [KWh/(m^2 year)]: best estimated EPI as output of the refined energy map workflow.

Real data of some building are listed in Table 1 shows, where hypothetic floor number was calculated by comparing the reference area and shape one.

The same results are reported in Fig. 5.

Some considerations on the validation process:

- For old buildings, TABULA overestimates the set of U_VALUES;
- The real energy certificates do not include information on the refurbishment level of the envelope.
- The Delta U Bridge for recent buildings is, in general, overestimated: around 10 % independently by the construction year.

Figure 6 represents the comparison of real and estimated data on the whole dataset composed by two cities, by year of construction.

The percentage gap between two values is much greater in the recent buildings, because uncertainties in glazing area are more relevant than U-values.

Table 3. Samples of validation result.

ID	Type	City	Floors	Year	Real EPi	Est. EPi	Fin. EPi
Building 1	SFH	TN	3	1934	394,29	390,38	390,38
Building 2	MFH	TN	3	1920	362,01	519,38	400,48
Building 3	MFH	TN	4	1990	256,97	301,8	259,56
Building 4	MFH	TN	2	1961	230,3	342,22	238,12
Building 5	MFH	CL	2	1973	82,26	128,44	76,88
Building 6	SFH	TN	1	2004	89,32	89,95	89,95
Building 7	SFH	TN	1	2012	94,28	112,03	112,03
Building 8	SFH	CL	4	2007	162,27	130,09	130,09
Building 9	MFH	TN	1	2012	42,92	128	53,56
Building 10	MFH	CL	2	2012	56,01	96,4	59,04
Block 1	AB	TN	4	1992	73,4	68,63	68,63
Block 2	AB	TN	5	2004	87,41	94,64	94,64
Block 3	AB	TN	1	1950	191,21	232,92	232,92
Block 4	AB	TN	1	1950	314,29	282,14	282,14

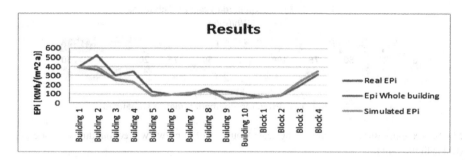

Fig. 5. Validation against buildings reported in Table 1.

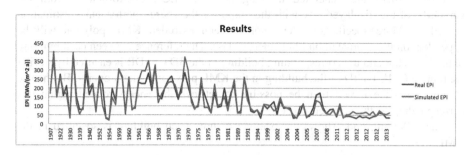

Fig. 6. Validation graph against the entire dataset.

The following diagram represents the normal distribution of absolute errors. As it is possible to see, the average error between estimated and real data is near the 21 % but, taking into account the previous considerations on the validation process, it is easy to understand that the error factor can be effectively decreased (Fig. 7).

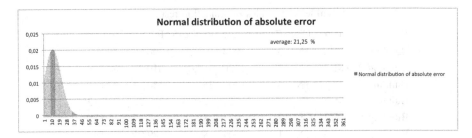

Fig. 7. Errors distribution against real and estimated data.

3.4 Energy Map Visualization

As described in the previous section about system architecture, via the WebGL-enabled visualization the project stakeholders can easily discover, compare and perform statistics on the estimated energy map data by accessing to a classical HTML web page.

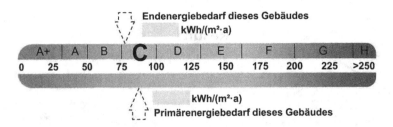

Fig. 8. Example Building color-coding scale [courtesy of http://www.greenbuildingadvisor.com] (Color figure online)

Energy maps are generated merging geometry LOD-1 information from the CityGML of the displayed city with the output of the energy performance estimation procedure. More specifically, the color of each extruded KML polygon will be dependent on the estimated building energy class. The reference between each building in the KML file and the corresponding building in the 3D CityDB is ensured by storing the unique GML UUID of the building in the KML polygon name property. By the use of a web service it will then be possible to retrieve the energy-related parameters corresponding to the selected object.

The following code shows an example of how each building is specified in the KML file.

```
<Style id="F">
   <PolyStyle>
     <color>FF0000FF</color>
     <fill>1</fill>
     <outline>0</outline>
   </PolyStyle>
 </Style>

<Placemark>
     <description>Building extruded test 1</description>
     <name>UUID_a2017297-d0cf-45ee-ae6d-94a5d4fcda03</name>
     <styleUrl>#F</styleUrl>
     <Polygon>
<altitudeMode>absolute</altitudeMode>
     <extrude>1</extrude>
     <outerBoundaryIs>
       <LinearRing>
         <coordinates>
             11.1263299929,46.0683712643,213.486
             11.1263070656,46.0684180254,214.313
             11.1262141309,46.0684011600,208.837
             11.1262401711,46.0683529640,213.74
             11.1263299929,46.0683712643,213.486
         </coordinates>
       </LinearRing>
     </outerBoundaryIs>
   </Polygon>
 </Placemark>
```

Referring to the code listed above, the first part is used to make a visual representation of the energy class determined by the estimation procedure. Figure 8 shows
the color coding of each building based on its estimated energy class. The second part
of the KML code is used to describe the extruded geometry of each building contained
in the source file.

Figure 9 shows, the energy map visualizer is composed by two interconnected
parts:

1. An HTML5 canvas that displays the WebGL virtual globe in which KML energy
   maps, based on CityGML LOD-1 geometries, are loaded;
2. A classical HTML tab, displaying the detailed energy data corresponding to the
   selected building. Comparisons between building energy efficiency characteristics
   can easily be performed using the "radar" diagram placed in the bottom-left part of
   the page. The diagram allows the comparison of the most important building proprieties between the current and the previously selected building.

Fig. 9. Energy map visualization.

3. For each involved building, the web client allows the possibility to refine the main algorithm input data, increasing the energy performance index estimation accuracy and providing some information for the apartment-level estimation (position of

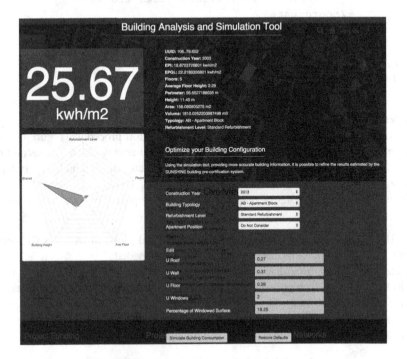

Fig. 10. Building analysis and simulation tool.

the apartment: ground, middle, last: already discussed in the previous chapter). Figure 10 shows how the web client allows stakeholders to use the building analysis and simulation tool in order to run the client-side EPI estimation.

4 Conclusion and Future Developments

In this paper we have presented some of the preliminary results of the SUNSHINE project. The use of a part of the TABULA building typology database with real building geometry information allows, for a large-scale, application of the building energy performance assessment and the underlying service-oriented system architecture supports a distributed access to the related services. Moreover, the use of the emerging WebGL technology ensures the largest available audience in terms of devices, both desktop and mobile, avoiding the development of device-dependent custom clients for 3D city map visualization.

Future developments will be linked to the estimation of energy performance data related to historical buildings. On the side of data structure and visualization, improvements will be focused on increasing the quality of the geometry displayed, making it possible to render buildings based on CityGML LoD-2 level of detail and on the development of more detailed building size type estimation procedures.

Acknowledgements. The project SUNSHINE has received funding from the EC, and it has been co-funded by the CIP-Pilot actions as part of the Competitiveness and innovation Framework Programme. The authors are solely responsible of this work, which does not represent the opinion of the EC. The EC is not responsible for any use that might be made of information contained in this paper.

References

1. Apache Community, Apache Tomcat. http://tomcat.apache.org/
2. Ballarini, I., et al.: Definition of building typologies for energy investigations on residential sector by TABULA IEE-project: application to Italian case studies. In: Proceedings of the 12th International Conference on Air Distribution in Rooms, Trondheim, Norway, pp. 19–22 (2012)
3. Bowerman, B., et al.: The vision of a smart city. 2nd International Life Extension Technology Workshop, Paris (2000)
4. Carrión, D., et al.: Estimation of the energetic rehabilitation state of buildings for the city of Berlin using a 3D city model represented in CityGML. In: ISPRS International Conference on 3D Geoinformation, p. 4 (2010)
5. Dalla Costa, S., et al.: A CityGML 3D geodatabase for buildings' energy efficiency. In: International Archives of the Photogrammetry, Remote Sensing and Spatial Information Sciences 38.4: C21 (2011)
6. Fenger, J.: Urban air quality. Atmos. Environ. **33**(29), 4877–4900 (1999)
7. Giffinger, R.: Smart cities: Ranking of European medium-sized cities. Final report, Centre of Regional Science, Vienna UT (2007)

8. Giffinger, R., Gudrun, H.: Smart cities ranking: an effective instrument for the positioning of the cities? ACE: Archit. City Environ. **4**(12), 7–26 (2010)
9. Gröger, G., et al.: OpenGIS city geography markup language (CityGML) encoding standard. Open Geospatial Consortium Inc. (2008)
10. Hay, G., et al.: HEAT - Home energy assessment technologies: a web 2.0 residential waste heat analysis using geobia and airborne thermal imagery. The International Archives of the Photogrammetry, Remote Sensing and Spatial Information Sciences, vol. XXXVIII-4/C7 (2010)
11. Kaden, R., Kolbe T.: City-wide total energy demand estimation of buildings using semantic 3D city models and statistical data. In: ISPRS Annals of the Photogrammetry, Remote Sensing and Spatial Information Sciences, vol. II-2/W1 (2013)
12. Krüger, A., Kolbe, T.: Building Analysis for urban energy planning using key indicators on virtual 3D city models - The energy atlas of Berlin. International Archives of the Photogrammetry, Remote Sensing and Spatial Information Sciences, Volume XXXIX-B2 (2012)
13. Loga, T.: Use of Building Typologies for Energy Performance Assessment of National Building Stocks: Existent Experiences in European Countries and Common Approach. IWU (2010)
14. Marrin, C.: WebGL specification. Khronos WebGL Working Group (2011)
15. Nouvel, R., et al.: CityGML-based 3D city model for energy diagnostics and urban energy policy support. In: Proceedings of BS2013, 13th Conference of International Building Performance Simulation Association, Chambéry, France (2013)
16. Special Interest Group 3D. http://www.sig3d.org/index.php?&language=en
17. Stadler, A., et al.: Making interoperability persistent: a 3D geo database based on CityGML. In: Lee, J., Zlatanova, S. (eds.) 3D Geo-Information Sciences, pp. 175–192. Springer, Heidelberg (2009)
18. Washburn, D., Sindhu, U.: Helping CIOs Understand "Smart City" Initiatives. Report by Forrester Research, Inc. (2009)
19. Wilson, T.: OGC® KML. OGC Encoding Standard, Version 2.0 (2008)

A Researcher's View on (Big) Data Analytics in Austria Results from an Online Survey

Ralf Bierig[1][(✉)], Allan Hanbury[1], Florina Piroi[1], Marita Haas[2],
Helmut Berger[2], Mihai Lupu[1], and Michael Dittenbach[2]

[1] Institute of Software Technology and Interactive Systems,
Vienna University of Technology, Favoritenstr. 9-11/188-1, 1040 Vienna, Austria
{bierig,hanbury,piroi,lupu}@ifs.tuwien.ac.at
[2] Max.recall Information Systems GmbH, Künstlergasse 11/1, 1150 Vienna, Austria
{m.haas,h.berger,m.dittenbach}@max-recall.com

Abstract. We present results from questionnaire data that were collected from leading data analytics researchers and experts across Austria. The online survey addresses very pressing questions in the area of (big) data analysis. Our findings provide valuable insights about what top Austrian data scientists think about data analytics, what they consider as important application areas that can benefit from big data and data processing, the challenges of the future and how soon these challenges will become relevant, and become potential research topics of tomorrow. We visualize results, summarize our findings and suggest a roadmap for future decision making.

Keywords: Data analysis · Data analytics · Big data · Questionnaire · Survey · Austria

1 Introduction

We are living in a data-saturated time. Continuous and large-scale methods for data analytics are needed as we now generate the impressive amount of 200 exabytes of data each year. This is equivalent to the volume of 20 million Libraries of Congress [1]. In 2012 each Internet minute has witnessed 100,000 tweets, 277,000 Facebook logins, 204 million email exchanges, and more than 2 million search queries fired to satisfy our increasing hunger for information [2].

This trend is accelerated technologically by devices that primarily generate digital data without the need for any intermediary step to first digitize analog data (e.g. digital cameras vs. film photography combined with scanning). Additional information is often automatically attached to the content (e.g. the exchangeable image file format 'Exif') that generates contextual metadata on a very fine-grained level. This means, when exchanging pictures, one also exchanges his or her travel destination, time (zone), specific camera configuration and the light conditions of the place, with more to come as devices evolve. Such sensors lead to a flood of machine-generated information that create a

© Springer International Publishing Switzerland 2015
M. Helfert et al. (Eds.): DATA 2014, CCIS 178, pp. 45–61, 2015.
DOI: 10.1007/978-3-319-25936-9_4

much higher spatial and temporal resolution than possible before. This 'Internet of Things' turns previously data-silent devices into autonomous hubs that collect, emit and process data at a scale that make it necessary to have automated information processing and analysis [1] to extract more value from data than possible with manual procedures. Today's enterprises are also increasing their data volumes. For example, energy providers now receive energy consumption readings from Smart Meters on a quarter-hour basis instead of once or twice per year. In hospitals it is becoming common to store multidimensional medical imaging instead of flat high-resolution images. Surveillance cameras and satellites are increasing in numbers and generate output with increasingly higher resolution and quality. Therefore, the focus today is on discovery, integration, consolidation, exploitation and analysis of this overwhelming data [1]. Paramount is the question of how all this (big) data should be analyzed and put to work. Collecting data is not an end but a means for doing something sensible and beneficial for the data owner, the business and the society at large. Technologies to harvest, store and process data efficiently have transformed our society and interesting questions and challenges have emerged of how society should handle these opportunities. While people are generally comfortable with storing large quantities of personal data remotely in a cloud there is also rising concern about data ownership, privacy and the dangers of data being intercepted and potentially misused [3].

In this paper, we present results of a study [4] that was conducted between June 2013 and January 2014 on the topic of (big) data analysis in Austria. Specifically, we present and highlight results obtained from an online survey that involved leading data scientists from Austrian companies and the public sector. The questionnaire was targeted to identify the status quo of (big) data analytics in Austria and the future challenges and developments in this area. We surveyed opinion from 56 experts and asked them about their understanding of data analytics and their projections on future developments and future research.

The paper first discusses related work and a status-quo of the potential application areas of (big) data in the next section. We then describe the method that was used for creating the questionnaire and for collecting and analyzing the feedback in Sect. 3. Results are presented and discussed in Sect. 4. In Sect. 5 we conclude and summarize our findings and suggest actions, in the form of a roadmap, that are based on our findings.

2   Related Work and State of (Big) Data Applications

Countries across the globe are eagerly developing strategies for dealing with big data. Prominent examples are the consultation process to create a Public-Private Partnership in Big Data currently underway in Europe,[1,2] work by the National Institute of Standards and Technology (NIST) Big Data Public Working Group[3]

---

[1] http://europa.eu/rapid/press-release_SPEECH-13-893_en.htm.

[2] All links have been accessed and validated in December 2014.

[3] http://bigdatawg.nist.gov.

as well as other groups [5] in the USA, and the creation of the Smart Data Innovation Lab[4] in Germany.

The recent and representative McKinsey report [6] estimates the potential global economic value of Big Data analytics between \$3.2 trillion to \$5.4 trillion every year. This value arises by intersecting open data with commercial data and thus providing more insights for customised products and services and enabling better decision making. The report identified the seven areas of education, transportation, consumer products, electricity, oil and gas, healthcare and consumer finance. We expanded this selection by specifically focusing on the Austrian market and its conditions before prompting participants with a comprehensive selection of application areas as described in Sect. 4.3. The remainder of this section will briefly review the scope of these application areas and why (big) data analytics could be interesting and helpful.

Healthcare has many stakeholders (i.e. patients, doctors, and pharmaceutical companies) each of which generate a wealth of data that often remains disconnected. In Austria there are also significant differences in how various federal states deal with patient data. There is great potential benefit in combining disconnected medical data sets for advancing healthcare, such as predictive modeling to improve drug development and personalized medicine to tailor treatments to individual genetic dispositions of patients [7, p. 37].

The *Energy and Utilities* sector now starts implementing the Smart Grid technology that allows recording energy consumption constantly every 15-minutes (i.e. Smart Metering) compared with manual monthly readings. This enables energy companies to better understand usage patterns and to adapt services and the energy grid to changing demands. Ecova[5], an energy and sustainablity management company, recently evaluated its energy data and published some interesting trends in consumer energy consumption [8]. They found that between 2008 and 2012, energy consumption across the US dropped by almost 9 % while water prices increased by almost 30 %. This demonstrates how (big) data analytics can help inform about large-scale changes in our society. There are however still many technical problems, primarily about privacy and security, and the issue that customers have only limited and slow access to their own consumption information. This were some of the reasons why Austria enabled consumers to opt-out from smart meters[6].

EScience is commonly seen as the extended and more data-intensive form of science supported by its increased capacity to generate and transform data [9]. An European Commision Report [10] describes the general infrastructure requirements, such as providing a generic set of tools for capture, validation, curation, analysis, and the archiving of data, the interoperability between different data sources, the (controlled and secured) access to data. Furthermore, incentives should be given to contribute data back to the infrastructure strengthened with

---

[4] http://www.sdil.de/de/.

[5] http://www.ecova.com.

[6] http://www.renewablesinternational.net/smart-meter-rollout-in-europe-rollback-in-austria/150/537/72823/

a range of financial models to ensure the sustainability of the infrastructure. A first basic e-Infrastructure is currently created in Austria to provide archiving services for scientific publications and data.[7]

Manufacturing and Logistics traditionally generates large data sets and requires sophisticated data analysis methods. This has developed further in recent years when moving to the fourth industrial revolution[8]. There are many applications of data analytics, such as Manufacturing Systems, Fault Detection, Decision Support, Engineering Design or Resource Planning, as described in [11]. These applications remain relevant and now further expand into big data applications that are more complex, more (socially) connected, and more interactive. The McKinsey report [6] foresees the (big) data applications in manufacturing to be in R&D, supply chain and in functions of production as a modern expansion of Harding's categories.

The sector of *Telecommunications* generally works with large data — as of May 2014, we have nearly 7 billion mobile subscriptions worldwide that represent 95.5 percent of the world population [12]. Each of these mobile phones transmit data every time we call, text, or access the internet. Moreover, mobile phones send passive information, such as when handing over between antennas and when estimating geo-positions. For this reason, telecommunications have long established big data at the core of their business model although they focus more on the real-time aspect of data and the customer as the target of their data analytics.

Transportation undergoes many changes as the economy moves into the information age where product-related services turn into information-related services. The growing amount of data, especially real-time data, needs to be managed and applied in such applications as price optimisation, personalisation, booking and travel management, customer relationship management [13]. In Austria, predictive data analysis is increasingly used for transportation incident management and traffic control, however at an initial stage.

Education can benefit from intelligent data analytics through Educational Data Mining [14] that develops methods for exploring data from educational settings to better understand students. These methods span across disciplines like machine learning, statistics, information visualization and computational modelling. There are initial efforts to include them in Moodle [15] in an attempt to make them broadly accessible.

Finance and Insurance deal with large data volumes of mostly transactional data from customers. The goal is to move such data from legacy storage-only solutions to distributed file systems, like Hadoop. This allows fraud detection [16] and custom-tailored product offers. An IBM study for example identified a 97 % increase in the number of financial companies that have gained competitive advantages with data analysis between 2010 and 2012 [17].

The gains for the *Public Sector and Government Administration* are mostly in tax and labour services and are benefitting both the people and the governments

---

[7] http://www.bibliothekstagung2013.at/doc/abstracts/Vortrag_Budroni.pdf.
[8] http://en.wikipedia.org/wiki/Industry_4.0.

that represent them. Citizens benefit from shorter paper trails (e.g. for re-entering data) and waiting times (e.g. for tax returns). At a European level, the public sector could reduce administrative costs by 15 % to 20 % and create an extra of €150 to €300 billion value according to [7]. Austria has taken essential steps for a more efficient public sector. The cities of Vienna, Linz and Graz for example provide much of their data as part of the Open Government Initiative[9] so that companies can develop applications that ease daily life. Electronic filing started in 2001 (ELAK-der elektronische Akt) to replace paper based filing and archiving in all Austrian ministries [18]. The system shortens the paper processing times by up to 15 % and already has over 9,500 users. However, these solutions often exist in isolation and there is much need to integrate these data silos and strengthen them with additional analytical capabilities.

While *Commerce and Retail* is traditionally involved in activities of business intelligence it now faces a scenario that shifts from the problem of actively collecting data to selecting the relevant from too much data available. One pioneer use of data analytics was the stock market where it made some a fortune and possibly also took part in stock market crashes [19]. Retail benefits significantly from (big) data analytics [7, p. 64–75] where customers have more options to compare and shop in a transparent form. Retailers make informed decisions by mining customer data (e.g. from loyalty cards and mobile applications).

Tourism and Hospitality deals with big data and can be an application for data analytics as tourists leave a rich digital trail of data. Those businesses who best anticipate the customers expectations will be able to sell more "experiences". In [20], an example is given of how tourism statistics combined with active digital footprints (e.g. social network postings) provide a representation of the tourist dispersion among sites in Austria.

Law and Law Enforcement deals with millions of cases and service calls every year. It was in Santa Cruz, California, where data from 5000 crimes was first applied to predict, and then prevent, upcoming offences. This might remind some of 'Minority Report' and Hunter S. Thompson would certainly describe it as a situation of "a closed society where everybody's guilty [and] the only crime is getting caught" [21, PartI,9]. However, the programme did, in a first round, reduce crime incidents by up to 11 % [22]. Financial crime detection also benefits from data analytics by processing and applying predictive methods on transactions [23]. In Austria concrete steps in this direction where taken by deciding that courts of law must use the electronic file system ELAK, as described in [18].

There are other areas that engage in (big) data analytics that have not been considered in the scope of this survey: In *Earth Observation*, many terabytes of data are generated and processed every day to forecast weather, predict earthquakes and estimate land use. *Agriculture* has drastically changed in the past half century by combining weather/environment data and data from machines and agricultural business processes for increased efficiency. In [7], *Media and Entertainment* is described as less capable for capturing the value of big data

---

[9] http://data.gv.at.

despite being IT intense. Big data can also be found in *Sports*, *Gaming*, and *Real Estate*. Let us not forget *Defence and Intelligence* — an area that most likely started the idea of collecting and correlating data from the masses.

Many other surveys have been conducted on the topic of big data and big data analytics by consulting companies, but these surveys usually concentrate on large enterprises.[10] A summary of the 2013 surveys is available [24]. A survey among people defining themselves to be Data Scientists has also been conducted to better define the role of Data Scientists [25]. Here, we consider the views of mostly academic scientists working in multiple areas related to data analytics, and hence we provide an unusual "academic" view of this emerging new field.

3   Method

Surveys are powerful tools when collecting opinion from the masses. Our main objective was to further specify our understanding of data analytics in Austria and to identify future challenges in this emerging field.

We followed the strategy of active sampling. The identification of Austrian stakeholders in data analytics formed the starting point: We first scanned and reviewed Austrian industry and research institutions based on their activities and research areas. We then identified key people from these institutions and asked about their opinions, attitudes, feedback and participation during a roadmapping process.

Our final contact list comprised 258 experts, all of them senior and visible data scientists, that we contacted twice and invited them to complete our questionnaire. This means our contact list has consensus-quality and represents the current situation and strength of senior data scientists in Austria. The survey was online between the beginning of September 2013 until the middle of October 2013. A total of 105 people followed the link to the survey resulting in a general response rate of 39 %. However, several of them turned down the questionnaire or cancelled their efforts after only one or two questions. We took a strict measure and removed those incomplete cases from the list of responses to increase the quality of the data. This reduced the original 105 responses (39 %) further down to 56 responses (21.7 %).

The general advantages of online surveys, such as truthfulness, increased statistical variation and improved possibilities for data analysis (e.g. [26,27]), unfortunately suffer from the problems of limited control, a higher demand on participants in terms of time and patience and the potential that people may be engaged in other, distracting activities that alter the results and increase the dropout rate (e.g. [28]). While our response rate of nearly 40 % is normal for

---

[10] Some examples: http://www-935.ibm.com/services/us/gbs/thoughtleadership/ibv-big-data-at-work.html, http://www.sas.com/resources/whitepaper/wp_58466.pdf, the Computing Research Association (CRA) http://www.cra.org/ccc/files/docs/init/bigdatawhitepaper.pdf and SAS http://www.sas.com/resources/whitepaper/wp_55536.pdf.

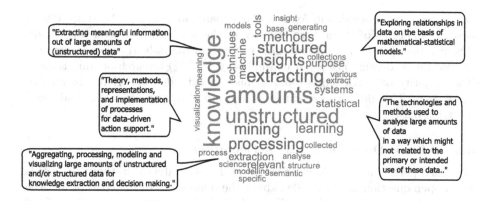

Fig. 1. What participants understood as data analytics.

online surveys [26], the high dropout rate in our specific case can be attributed to the complex nature of the subject.

The data of the survey was collected anonymously with LimeSurvey[11] and was analyzed with R, a statistical software package[12].

4 Survey Results and Discussion

This section highlights the results we obtained from the data and focuses on four areas. First, the demographic information about the participants (e.g. their age, area of work, and their work experience) that helps us to get a better understanding of the characteristics of a typical data scientist in Austria. Second, we look at the application areas of data analytics and how participants projected the future relevance of these areas. Third, we investigated the future challenges of data analytics. Fourth, we analysed free text submissions from questions about research priorities and the need for immediate funding to get a better understanding of possible future directions, interests and desires. We omitted replies for questions that only had a very limited text response that would not be meaningful to analyze statistically. The survey questions are provided in detail in the appendix of [29].

4.1 Participants: Our Data Scientists

The data presented in this paper is based on the opinions of 56 people who completed the questionnaire — four female (7.1 %) and 52 male (92.9 %). This gender distribution is similar in the original contact list — 26 female (10.1 %) and 232 male (89.9 %) — and therefore represents the current gender situation in

---

[11] http://www.limesurvey.org/de/.

[12] http://www.r-project.org/. The survey questions are provided in the appendix of [29].

the data science profession in Austria. Participants were mostly Austrians (96 %) and the majority of them were working in the research and academic sector. About a fifth (21.4 %) of all responses came from the industry. The larger part worked for academic (55.4 %) or non-industry (33.9 %) research organisations.[13] The majority of participants (80.3 %) had an extended experience of nine or more years. This defines our sample as a group of mostly academic, male, and Austrian data scientists.

4.2   What is Data Analytics? A Definition

We asked participants to describe the term 'Data Analytics' in their own words as an open question to get an idea about the dimensions of the concept and the individual views on the subject. Figure 1 depicts a summary word cloud from the collected free-text responses for all those terms that repeatedly appeared in the response.[14] It further depicts a small set of representative extracts from the comments and definitions that participants submitted. Overall, the comments were very much focused on the issue of large data volumes, the process of knowledge extraction with specific methods and algorithms and the aggregation and combination of data in order to get new insights. Often it was related to machine learning and data mining but as a wider and more integrative approach. Only very few respondents labeled Data Analytics to be simply a modern and fashionable word for data mining or pattern recognition.

4.3   Important Application Areas

Based on the literature review that preceded this survey, we identified the main application areas of data analytics in Austria as healthcare, commerce, manufacturing and logistics, transportation, energy and utilities, the public sector and the government, education, tourism, telecommunication, e-science, law enforcement, and finance and insurance. Figure 2 shows the relative importance of these areas as attributed by participants. Selections were made in binary form with multiple selections possible. The figure shows that the area of healthcare is perceived as a strong sector for data analytics (66.1 %) followed by energy (53.6 %), manufacturing and e-science (both 50.0 %). As a sector that is perceived to benefit only little from (big) data analytics are tourism and commerce (both 23.2 %). This is despite the fact that these areas are large in Austria based on demographic data as provided by Statistics Austria.[15]

---

[13] Multiple selections were possible which means that these numbers do not add up to 100 %.

[14] We only included terms that appeared a least three times and we filtered with an english and a topical stop word list (e.g. terms like 'and' or 'etc' and terms like 'data' or 'analytics').

[15] In 2010, 19.3 % of the employed worked in commerce and 9.1 % in the gastronomical and leisure sector (source: 'Ergebnisse im Ueberblick: Statistik zur Unternehmensdemografie 2004 bis 2010', available at http://www.statistik.at/web_de/statistiken/unternehmen_arbeitsstaetten/arbeitgeberunternehmensdemografie/index.html, extracted 09-12-2014.

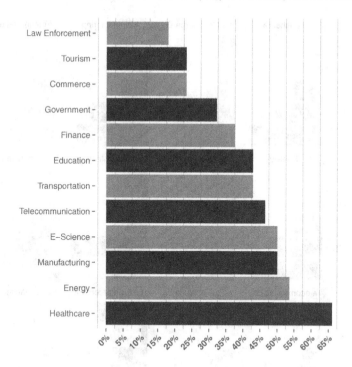

Fig. 2. Important application areas for data analytics.

We additionally asked participants to provide future projections for these application areas w.r.t. how they think these areas would become important for data analytics in the future (see Fig. 3). Here, participants rated the application areas based on their relevance for the *short, middle* and *long term* future. The diagram also visualizes the amount of uncertainty in these projections as participants could select if they were unsure or even declare an application area as unimportant. The figure shows that application areas that are perceived as strong candidates (e.g. healthcare, energy and telecommunication) are all marked as relevant for the short term future with decreasing ratings on the longer timeline. Less strongly perceived application areas, such as law enforcement and tourism have results that are less clearly expressed with a stronger emphasis on a longer time frame. The amount of uncertainty about these areas is also much higher. Law enforcement is perceived as both less important and not benefiting from data analytics. This is conceivable as law enforcement may not be perceived as an independent sector, as this is the case in the United States [30] where data analytics already assists the crime prediction process with data mining, e.g. with the use of clustering, classification, deviation detection, social network analysis, entity extraction, association rule mining, string comparison and sequential pattern mining. It comes as a surprise that tourism in Austria was both perceived as rather unimportant and also as an area that would only benefit from data analytics in the mid- and long-term future. The large proportion of uncertainty shows that experts seem to be rather unsure about the future of these two sectors.

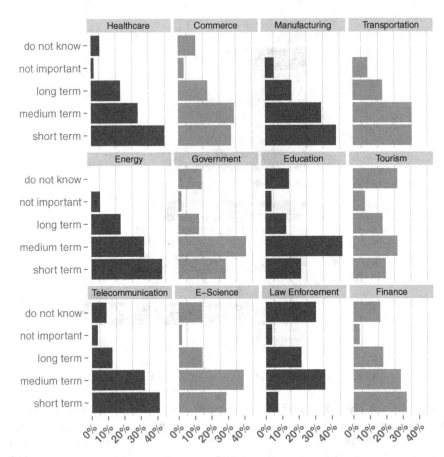

Fig. 3. Application areas where data analytics will become important in the short, middle and long term.

4.4   Current and Future Challenges

Based on the literature review, we identified the main challenges of data analytics in the areas of privacy and security, algorithms and scalability, getting qualified personnel, the preservation and curation process, the evaluation and benchmarking, and data ownership and open data. We now asked participants to categorize these challenges into three groups: *Short term* if they see it as an issue of the very near future, *medium term* if there is still time and *long term* if this might become an issue some time in the far future. Our intent was to obtain a priority that can help us to identify possible actions and recommendations for decision making. Figure 4 depicts all responses for all categories and also includes the amount of uncertainty (*do not know*) and how unimportant people thought it to be (*not important*).

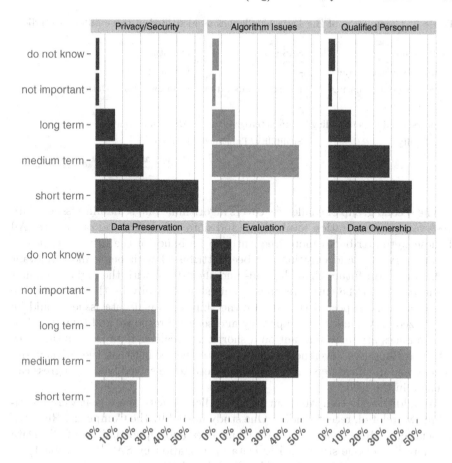

Fig. 4. Future challenges in data analytics for the short, middle and long term.

All challenges share that they are perceived as all being relevant in the short and middle-term future, with high certainty throughout. Upon closer investigation, the response can then be further divided into two groups.

The first group of challenges consists of privacy and security and the issue of qualified personnel. These issues are perceived as considerably more important and pressing in the short term future than the middle and the long term future. It was especially striking how important *privacy and security* was emphasised throughout the entire study — including this online questionnaire. This strong impact might be partially attributed to the very recent NSA scandal. However, it might also be that our target group quite naturally possesses a heightened sensitivity to the potential dangers of (big) data analytics and the often unprotected flow of personal data on the open web. *Qualified personnel* is a problem of the near future and has been discussed in the literature throughout many studies. This is well confirmed in our own findings and an important issue to address in future decision making.

Table 1. Comparison of Big Data Challenges as identified in three different studies.

CRA study	SAS study	Our study
Lack of skills/Experience	-	Personnel
Accessing data/Sharing	Human collaboration	Data ownership/Data preservation
Effective use	-	-
Analysis and understanding	Data heterogenity	Algorithm issues
Scalibility	Scalibility/Timeliness	Evaluation
-	Privacy	Privacy and security

The second group of challenges covers algorithmic and scalability issues, data preservation and curation, evaluation, and data ownership and open data. All of these were attributed more frequently to be issues of the future. Ironically, data preservation and curation has been attributed with being more relevant in the long-term future than the mid- and short-term with the highest amount of uncertainty in the entire response. This should ideally be the opposite. We would have also expected that data ownership and open data issues would be categorized very similar to the privacy and security response and that the algorithmic issues are more relevant on a short term scale as data is mounting very fast. The responses nevertheless demonstrate the feeling that the privacy and security and qualified personnel challenges need to be solved before progress can be made in the field.

We additionally compared our list of challenges with those that were identified in two related, recent studies: One study hosted by the Computing Research Association (CRA) that focused on Challenges and Opportunities of Big Data in general[16] and one study on Big Data visualization by SAS.[17] In Table 1, we refer to them as 'CRA Study', 'SAS Study' and 'Our Study' and compare six challenge categories that were identified across these studies. The challenges are presented in no particular order, however, the reader can compare challenge categories horizontally in the table. A dash (-) means that a particular challenge was not identified by a study. We related the categories with each other to give the reader an overview about the similarities and differences from three perspectives. Naturally, the categories did not always represent a perfect match. For example, the challenge of data access and data sharing was addressed as the need for human collaboration in the SAS study and our own study identified the challenge of data ownership and the challenge of preserving data in this category. However, the issue of a lack of personnel was also identified as a lack of skills and experience in the CRA study. Whereas privacy was clearly addressed in the SAS study and more comprehensively combined with security in our study, the CRA study did not consider it a challenge at all. Overall, this comparison shows that there is considerable agreement between studies with respect to future

---

[16] http://www.cra.org/ccc/files/docs/init/bigdatawhitepaper.pdf.

[17] http://www.sas.com/resources/whitepaper/wp_55536.pdf.

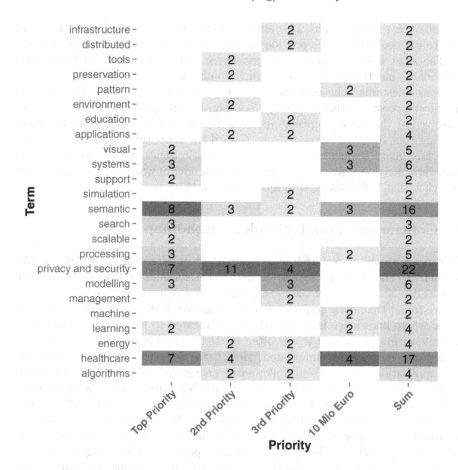

Fig. 5. Priorities for future research as expressed by participants.

challenges. It would be interesting to further extend this comparison to a much wider range of studies in future work.

4.5 Future Research Topics

We prompted participants with two questions about future research in (big) data analytics. The first question asked them to enter free text on topics of their preferred future research which they had to prioritise by three levels (top priority, 2nd priority, 3rd priority). The second question can be seen as a refinement of the top priority level of the previous question and asked them to describe which research topic they would like to see publicly funded with 10 Million €. Again, this was submitted as free text allowing participants to contribute their ideas in a completely free and unrestricted form. Figure 5 shows the term frequencies of those texts for all three priorities and also the text for the

10 Million € research topic. This allows for easy comparisons. A sum across all four columns of frequencies provides an overview about the entire topic space.

The most frequent themes are privacy/security (mentioned 22 times) and healthcare (mentioned 17 times) which coincides with the findings from the previous questions. The importance of privacy and security was found to be the most pressing future challenge (see Fig. 4). Likewise, healthcare was also perceived as the most promising application domain (see Fig. 2) with the most pressing time line that strongly leans toward the short-term future (see Fig. 3). The third most frequent keyword were semantic issues (mentioned 16 times) that were more extensively investigated in a number of workshops and an expert interview that is documented in more detail in [4].

5   Conclusions and Technology Roadmap

This paper presented results from a study on (big) data analysis in Austria. We summarized the opinions of Austrian data scientists sampled from both industry and academia and some of the most pressing and current issues in the field.

We found that data analytics is understood as dealing with large data volumes, where knowledge is extracted and aggregated to lead to new insights. It was interesting to see that it was often related to data mining but viewed more widely and more highly integrated. Healthcare was seen as the most important application area (66.1 %), followed by energy (53.6 %), manufacturing and logistics (50.0 %) and e-science (50.0 %) with big potential in the short term future. Other areas were judged less important, such as tourism (23.2 %) and law enforcement (17.9 %), with high uncertainty. We found that our literature-informed list of challenges were confirmed by our respondents, however only privacy/security and the challenge to get qualified personnel was strongly attributed to the very near future. Algorithm issues, data preservation, evaluation and data ownership were seen as challenges that become more relevant only in the longer run. Research priorities and funding requests where strongly targeted to privacy/security (mentioned 22 times), healthcare (mentioned 17 times) and semantic issues (mentioned 16 times). This result conforms largely to the findings in the other parts of the study.

Based on the results of the survey presented in this paper, along with the outcomes of three workshops and interviews, a technology roadmap consisting of a number of objectives was drawn up. The roadmap actions are described in much greater detail in [4] as part of the complete report that focuses on all parts of the study. This is outside the scope of this paper that focuses on the details of the online survey. In summary, the identified challenges, together with their careful evaluation, have led to three categories of actions that are manifested in this roadmap.

First, to meet the challenges of data sharing, evaluation and data preservation, an objective in the roadmap is to create a "Data-Services Ecosystem" in Austria. This is related to an objective to create a legal and regulatory framework that covers issues such as privacy, security and data ownership, as such a framework is necessary to have a functioning Ecosystem. In particular, it is suggested to fund a study project to develop the concept of such an Ecosystem, launch measures to educate and encourage data owners to make their data and problems available, and progress to lighthouse projects to implement and refine the Ecosystem and its corresponding infrastructure. Furthermore, it is recommended to develop a legal framework and create technological framework controls to address the pressing challenges of privacy and security in data analytics.

Second, technical objectives are to overcome challenges related to data integration and fusion and algorithmic efficiency, as well as to create actionable information and revolutionise the way that knowledge work is done. We suggest to fund research that focuses on future data preservation, to develop fusion approaches for very large amounts of data, to create methods that assure anonymity when combining data from many sources, to enable real time processing, and to launch algorithmic challenges. A full list of suggestions are described in more detail in [4].

The third and final objective in the roadmap is to increase the number of data scientists being trained. We suggest a comprehensive approach to create these human resources and competences through educational measures at all levels: from schools through universities and universities of applied sciences to companies. The issue of having more and highly skilled data scientists soon is an issue that requires immediate action to secure the future prosperity of the Austrian (big) data analytics landscape.

Acknowledgements. This study was commissioned and funded by the Austrian Research Promotion Agency (FFG) and the Austrian Federal Ministry for Transport, Innovation and Technology (BMVIT) as FFG ICT of the Future project number 840200. We thank Andreas Rauber for his valuable input. Information about the project and access to all deliverables are provided at http://www.conqueringdata.com.

References

1. Dandawate, Y. ed.: Big Data: Challenges and Opportunities. Infosys Labs Briefings., vol. 11, Infosys Labs (2013). http://www.infosys.com/infosys-labs/publications/Documents/bigdata-challenges-opportunities.pdf. Last visited, December 2014
2. Temple, K.: What Happens in an Internet Minute? (2013). http://scoop.intel.com/what-happens-in-an-internet-minute/. Last visited, December 2014
3. Boyd, D., Crawford, K.: Critical questions for big data. Inf. Commun. Soc. **15**, 662–679 (2012)
4. Berger, H., Dittenbach, M., Haas, M., Bierig, R., Hanbury, A., Lupu, M., Piroi, F.: Conquering Data in Austria. bmvit (Bundesministerium für Verkehr, Innovation and Technology, Vienna, Austria (2014)
5. Agrawal, D., Bernstein, P., Bertino, E., Davidson, S., Dayal, U., Franklin, M., Gehrke, J., Haas, L., Halevy, A., Han, J., Jagadish, H.V., Labrinidis, A., Madden, S., Papakonstantinou, Y., Patel, J. M., Ramakrishnan, R., Ross, K., Shahabi, C., Suciu, D., Vaithyanathan, S., Widom, J.: Challenges and Opportunities with Big Data (2012). http://www.cra.org/ccc/files/docs/init/bigdatawhitepaper.pdf. Last visited, December 2014

6. Manyika, J., Chui, M., Groves, P., Farrell, D., Kuiken, S.V., Doshi, E.A.: Open Data: Unlocking Innovation and Performance with Liquid Information. McKinsey Global Institute, Kenya (2013)

7. Manyika, J., Chui, M., Brown, B., Bughin, J., Dobbs, R., Roxburgh, C., Byers, A.H.: Big Data: The Next Frontier for Innovation, Competition, and Productivity. McKinsey Global Institute, Kenya (2011)

8. Hardesty, L.: Ecova customers cut electric consumption intensity 8.8%, shows study (2013). http://www.energymanagertoday.com/ecova-customers-cut-electric-consumption-intensity-8-8-shows-study-093633/. Last visited, December 2014

9. Harvard Business Review: Data scientist: The sexiest job of the 21st century (2012). http://hbr.org/2012/10/data-scientist-the-sexiest-job-of-the-21st-century/. Last visited, December 2014

10. Riding the Wave: How Europe can gain from the rising tide of scientific data. European Commission (2010)

11. Harding, J.A., Shahbaz, M., Kusiak, A.: Data mining in manufacturing. Rev. J. Manuf. Sci. Eng. **128**(4), 969–976 (2006)

12. ICT: The world in 2014: ICT facts and figures (2014)

13. Ellis, S.: Big Data and Analytics Focus in the Travel and Transportation Industry (2012). http://h20195.www2.hp.com/V2/GetPDF.aspx%2F4AA4-3942ENW.pdf. Last visited, December 2014

14. Baker, R., Yacef, K.: The state of educational data mining in 2009 a review and future visions. J. Educ. Data Min. J. Educ. Data Min. **1**, 3–17 (2009)

15. Romero, C., Espejo, P., Zafra, A., Romero, J., Ventura, S.: Web usage mining for predicting final marks of students that use Moodle courses. Comput. Appl. Eng. Educ. **21**, 135–146 (2013)

16. The Economist: Big data: Crunching the numbers (2012). http://www.economist.com/node/21554743. Last visited, December 2014

17. Turner, D., Schroeck, M., Shockley, R.: Analytics: The real-world use of Big Data in financial services. IBM Global Business Services, Executive report (2013)

18. Müller, H.: ELAK, the e-filing system of the Austrian Federal Ministries (2008). http://www.epractice.eu/en/cases/elak. Last visited, November 2013

19. Findings regarding the market events of may 6, 2010. U.S. Securities and Exchange Commission and the Commodity Futures Trading Commission (2010). http://www.sec.gov/news/studies/2010/marketevents-report.pdf. Last visited, December 2014

20. Koerbitz, W., Önder, I., Hubmann-Haidvogel, A.: Identifying tourist dispersion in austria by digital footprints. In: Cantoni, L., Xiang, Z.P. (eds.) Information and Communication Technologies in Tourism 2013, pp. 495–506. Springer, Heidelberg (2013)

21. Thompson, H.S.: Fear and Loathing in Las Vegas: A Savage Journey to the Heart of the American Dream. Modern Library, Vintage Books (1971)

22. Olesker, A.: White Paper: Big Data Solutions For Law Enforcement, IDC White paper (2012)

23. Mehmet, M., Wijesekera, D.: Data analytics to detect evolving money laundering 71–78. In: Laskey, K.B., Emmons, I., Costa, P.C.G. (eds.) Proceedings of the Eighth Conference on Semantic Technologies for Intelligence, Defense, and Security, STIDS 2013, CEUR Workshop Proceedings, vol. 1097, pp. 71–78 (2013)

24. Press, G.: The state of big data: What the surveys say (2013). http://www.forbes.com/sites/gilpress/2013/11/30/the-state-of-big-data-what-the-surveys-say/. Last visited, December 2014

25. Harris, H., Murphy, S., Vaisman, M.: Analyzing the Analyzers: An Introspective Survey of Data Scientists and Their Work. O'Reilly, USA (2013)
26. Batinic, B.: Internetbasierte befragungsverfahren. Österreichische Zeitschrift für Soziologie **28**, 6–18 (2003)
27. Döring, N.: Sozialpsychologie des Internet. Die Bedeutung des Internet für Kommunikationsprozesse, Identitäten, soziale Beziehungen und Gruppen. 2nd edn. Hogrefe, Göttingen (2003)
28. Birnbaum, M.H.: Human research and data collection via the internet. Annu. Rev. Psychol. **55**, 803–832 (2004)
29. Bierig, R., Hanbury, A., Haas, M., Piroi, F., Berger, H., Lupu, M., Dittenbach, M.: A glimpse into the state and future of (big) data analytics in austria - results from an online survey. In: DATA 2014 - Proceedings of 3rd International Conference on Data Management Technologies and Applications, Vienna, Austria, 29–31 August, 2014, pp. 178–188 (2014)
30. Norton, A.: Predictive policing - the future of law enforcement in the trinidad and tobago police service. Int. J. Comput. Appl. **62**, 32–36 (2013)

Accurate Data Cleansing through Model Checking and Machine Learning Techniques

Roberto Boselli, Mirko Cesarini$^{(\boxtimes)}$, Fabio Mercorio$^{(\boxtimes)}$,
and Mario Mezzanzanica

Department of Statistics and Quantitative Methods - CRISP Research Centre,
University of Milan-Bicocca, Milan, Italy
{roberto.boselli,mirko.cesarini,fabio.mercorio,
mario.mezzanzanica}@unimib.it

Abstract. Most researchers agree that the quality of real-life data archives is often very poor, and this makes the definition and realisation of automatic techniques for cleansing data a relevant issue. In such a scenario, the Universal Cleansing framework has recently been proposed to automatically identify the most accurate cleansing alternatives among those synthesised through model-checking techniques. However, the identification of some values of the cleansed instances still relies on the rules defined by domain-experts and common practice, due to the difficulty to automatically derive them (e.g. the date value of an event to be added).

In this paper we extend this framework by including well-known machine learning algorithms - trained on the data recognised as consistent - to identify the information that the model based cleanser couldn't produce. The proposed framework has been implemented and successfully evaluated on a real dataset describing the working careers of a population.

Keywords: Data cleansing · Model based data quality · Machine learning · Labour market data

1 Introduction

Data quality improvement techniques have rapidly become an essential part of the KDD process - Knowledge Discovery in Databases (KDD) [18] - as most researchers agree that the quality of data is frequently very poor, and this makes data quality improvement and analysis crucial tasks for guaranteeing the believability of the overall KDD process[1] [25,31,40]. In such a direction, the *data cleansing* (a.k.a. data cleaning) research area concerns with the identification of a set of domain-dependent activities able to cleanse a dirty database (with respect to some given quality dimensions).

---

[1] Here the term *believability* is intended as "the extent to which data are accepted or regarded as true, real and credible" [48].

© Springer International Publishing Switzerland 2015
M. Helfert et al. (Eds.): DATA 2014, CCIS 178, pp. 62–80, 2015.
DOI: 10.1007/978-3-319-25936-9_5

Although a lot of techniques and tools are available for cleansing data e.g., the ETL tools (Extract, Transform and Load, which cover the data preprocessing and transformation tasks in the KDD process [18]), a quite relevant amount of data quality analysis and cleansing design still relies on the experience of domain-experts that have to specify ad-hoc data cleansing procedures [46].

A relevant issue is how to support the development of cleansing procedures being accurate or (at least) making sense for the analysis purposes. Focusing on model based data cleansing, [23] has clarified that weakly-structured data can be modelled as a random walk on a graph. The works [34,37] describe the use of graph-search algorithms for both verifying the data quality and identifying cleansing alternatives, the latter problem was recently expressed also in terms of AI Planning problem [6,8]. The Universal Cleanser described in [34] requires a policy to select a cleansing action when several ones are available (since a data inconsistency can be fixed in several ways, all of them compliant with the data domain model). [35] shows how to automatically identify a policy for producing accurate results. This paper extends the work of [34,35] by using a machine learning approach to improve data cleansing activities.

2 Motivation and Contribution

The *longitudinal data* (i.e., repeated observations of a given subject, object or phenomena at distinct time points) have received much attention from several academic research communities among the time-related data e.g. [15,22,23,26, 30,45].

A strong relationship exists between *weakly-structured* and time-related data. Namely, let $Y(t)$ be an ordered sequence of observed data (e.g., subject data sampled at different time $t \in T$), the observed data $Y(t)$ are weakly-structured if and only if the trajectory of $Y(t)$ resembles a random walk on a graph [14,27].

Then, the Universal Cleansing framework introduced in [34] can be used as a model-based approach to identify all the alternative cleansing actions. Namely, a graph modelling a weakly-structured data(set) can be used to derive a Universal Cleanser i.e., a taxonomy of possible errors (the term error is used in the sense of inconsistent data) and provides the set of possible corrections. The adjective *universal* means that (the universe of) every feasible inconsistency and cleansing action is identified with respect to the given model.

[35] focused on the identification of a policy driving the choice of which correction has to be applied when severals are available. The policy is inferred by evaluating the possible cleansing results using an accuracy like metric over an artificially soiled dataset. This paper shows how the cleansing framework described in [34,35] can be enhanced by using a machine learning approach to precisely identify the data values that cannot be completely guessed using the model e.g., a model describing a dataset expected behaviour can be used to identify a missing event, to guess the event type, and to provide some constraints about the event date (the timeslot pointed out by the supposed previous and subsequent events), but cannot precisely identify the event date. The following example should help in clarifying the matter.

Motivating Example. Some interesting information about the population can be derived using the labour administrative archive, an archive managed by a European Country Public Administration that records job start, cessation, conversion and extension events. The data are collected for administrative purposes and used by different public administrations.

An example of the labour archive content is reported in Table 1, a data quality examination on the labour data is shortly introduced. Data consistency rules can be inferred by the country law and common practice: an employee having one full-time contract can have no part-time contracts at the same time, an employee can't have more than k part-time contract at the same time (we assume $k = 2$). This consistency behaviour can be modelled through the graph shown in Fig. 1. Consistent data can be described by a path on the graph, while a path does not exist for inconsistent data.

The reader would notice that the working history reported in Table 1 is inconsistent with respect to the semantic just introduced: a full-time Cessation is missing for the contract started by *event 03*. Looking at the graph, the worker was in node emp_{FT1} when *event 04* was received, however, *event 04* brings nowhere from node emp_{FT1}. Several cleansing options are available:

Table 1. An example of data describing the working career of a person. *P.T.* is for part-time, *F.T.* is for full-time.

Event #	Event Type	Employer-ID	Date
01	P.T. Start	Firm 1	$12^{th}/01/2010$
02	P.T. Cessation	Firm 1	$31^{st}/03/2011$
03	F.T. Start	Firm 2	$1^{st}/04/2011$
04	P.T. Start	Firm 3	$1^{st}/10/2012$
05	P.T. Cessation	Firm 3	$1^{st}/06/2013$
...	...	...	...

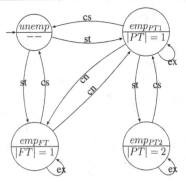

Fig. 1. A graph representing the dynamics of job careers where $st = start$, $cs = cessation$, $cn = conversion$, and $ex = extension$. The PT and FT variables count respectively the number of part-time and the number of full-time jobs.

(1) A typical cleansing approach would add a full-time cessation event in a date between *event 03* and *event 04* (this would allow one to reach the node *unemp* through the added event and then the node *emp$_{PT1}$* via *event 04*);

(2) to add a conversion event from full-time to part-time between *event 03* and *event 04* (reaching node *emp$_{PT1}$*).

Then, several questions arise: how to choose between these two alternatives? How to identify the dates of the events to be added?

In such a scenario, one might look at the *consistent* careers of the dataset where similar situations occurred in order to derive a criterion, namely a policy, for (1) selecting the most accurate cleansing option with respect to the dataset to be cleansed and (2) to identify the (missing event) information like the dates.

Contribution. This paper contributes to the problem of developing cleansing solutions for improving data quality. To this end, we extend a framework, built on top of the Universal Cleansing approach [34,35], that allows one:

(1) to model complex data quality constraints over longitudinal data, the former representing a challenging issue. Indeed, as [13] argue, the consistency requirements are usually defined on either (i) a single tuple, (ii) two tuples or (iii) a set of tuples. While the first two classes can be modelled through Functional Dependencies and their variants, the latter class requires reasoning with a (finite but not bounded) set of data items over time as the case of longitudinal data, and this makes the exploration-based techniques a good candidate for that task;

(2) to automatically generate *accurate* cleansing policies with respect to the dataset to be analysed. Specifically, an accuracy based distance is used to identify the policy that can better restore to its original version an artificially soiled dataset. Clearly, several policy selection metrics can be used and consequently the analysts can evaluate several cleansing opportunities;

(3) to precisely identify the data values that cannot be completely guessed using the model based cleansing approach;

(4) to apply the proposed framework on a real-life benchmark dataset modelling Labour Domain Data.

The outline of this paper is as follows. Section 3 outlines the Universal Cleansing Framework, Sect. 4 describes the automatic policy identification process, Sect. 5 shows how machine learning can be used to further refine the cleansing results of the output of the cleanser developed using the Universal Cleansing Framework. Section 6 shows the experimental results, Sect. 7 surveys the related works, and finally Sect. 8 draws the conclusions and outlines the future work.

3   Universal Cleansing

Here we briefly summarise the key elements of the framework described in [34,35] for evaluating and cleansing longitudinal data sequences.

Intuitively, let us consider an events sequence $\epsilon = e_1, e_2, \ldots, e_n$ modelling the working example of Table 1. Each event e_i will contain a number of observation variables whose evaluation determines a snapshot of the subject's *state*[2] at time point i, namely s_i. Then, the evaluation of any further event e_{i+1} might change the value of one or more state variables of s_i, generating a new state s_{i+1} in such a case. More formally we have the following.

Definition 1 (Events Sequence). *Let $\mathcal{R} = (R_1, \ldots, R_l)$ be the schema of a database relation. Then,*

(i) *An event $e = (r_1, \ldots, r_m)$ is a record of the projection $(R_1, \ldots, R_m)$ over $\mathcal{Q} \subseteq \mathcal{R}$ with $m \leq l$ s.t. $r_1 \in R_1, \ldots, r_m \in R_m$;*

(ii) *Let $\sim$ be a total order relation over events, an event sequence is a $\sim$-ordered sequence of events $\epsilon = e_1, \ldots, e_k$.*

A Finite State Event Dataset *(FSED) S_i is an event sequence while a* Finite State Event Database *(FSEDB) is a database S whose content is $S = \bigcup_{i=1}^{k} S_i$ where $k \geq 1$.*

3.1 Data Consistency Check

According to the approach described in [34] an expected data behaviour is encoded on a transition system (the so-called *consistency model*) so that each data sequence can be represented as a pathway on a graph. The transition systems can be viewed as a graph describing the consistent evolution of weakly-structured data. Indeed, the use of a sequence of events $\epsilon = e_1, e_2, \ldots, e_n$ as input actions of the transition system deterministically determines a path $\pi = s_1 e_1 \ldots s_n e_n s_{n+1}$ on it (i.e., a *trajectory*), where a state s_j is the state resulting after the application of event e_i on s_i. This allows authors to cast the data consistency verification problem into a model checking problem, which is well-suited for performing efficient graph explorations.

Namely, a model checker generates a trajectory for each event sequence: if a violation has been found, both the trajectory and the event that triggered the violation are returned, otherwise the event sequence is consistent. More details can be found in [32, 35].

Generally speaking, such a consistency check can be formalised as follows.

Definition 2 (ccheck). *Let $\epsilon = e_1, \ldots, e_n$ be a sequence of events according to Definition 1, then ccheck : $FSED \to \mathbb{N} \times \mathbb{N}$ returns the pair $< i, er >$ where:*

(1) *i is the index of a minimal subsequence $\epsilon_i = e_1, \ldots, e_i$ such that ϵ_{i+1} is inconsistent while $\forall j : j \leq i \leq n$, the subsequences ϵ_j are consistent.*

(2) *er is zero if ϵ_n is consistent, otherwise it is a natural number which uniquely describes the inconsistency error code of the sequence ϵ_{i+1}.*

---

[2] The term "state" here is considered in terms of a value assignment to a set of finite-domain state variables.

By abuse of notation, we denote as $first\{ccheck(S)\}$ the index i of the pair whilst $second\{ccheck(S)\}$ denotes the element er.

The *ccheck* function was implemented through model-checking based techniques and also used for the (Multidimensional) Robust Data Quality Analysis [5].

The consistency requirements over longitudinal data (e.g. the example presented in this paper) could be expressed by means of Functional Dependencies (FDs). However, as illustrated in [35], using functional dependencies presents several drawbacks that we summarize as follows: (1) an expensive join of k relations is required to check constraints on sequences of k elements; (2) the start and cessation of a Part Time (e.g. events 01 and 02) can occur arbitrarily many times in a career before the event 04 is met (where the inconsistency is detected), a k high enough can be identified for specific cases, however (if for any reason) k should be modified (e.g., a greater value is required), also the FDs set should be extensively changed; (3) the higher the value of k, the more the set of possible sequence (variations) to be checked; (4) Functional Dependencies do not provide any hint on how to fix inconsistencies.

3.2   Universal Cleanser (UC)

When an inconsistency is found, the framework performs a further graph exploration for generating corrective events.

To clarify the matter, let us consider again an inconsistent event sequence $\epsilon = e_1, \ldots, e_n$ as the corresponding trajectory $\pi = s_1 e_1 \ldots s_n e_n s_{n+1}$ presents an event e_i leading to an inconsistent state s_j when applied on a state s_i. What does "cleansing" such an event sequence mean? Intuitively, a *cleansing event sequence* can be seen as an alternative trajectory on the graph leading the subject's state from s_i to a new state where the event e_i can be applied (without violating the consistency rules). In other words, a *cleansing event sequence* for an inconsistent event e_i is a sequence of *generated* events that, starting from s_i, makes the subject's state able to reach a new state on which the event e_i can be applied resulting in a consistent state (i.e., a state that does not violate any consistency property). For example, considering the inconsistency described in Sect. 1 and the graph shown in Fig. 1, such cleansing action sequences are paths from the node emp_{FT} to the node emp_{PT1} or to the node *unemp*. For our purposes, we can formalise this concept as follows.

Definition 3 (Cleansing Event Sequence). *Let $\epsilon = e_1, \ldots, e_n$ be an event sequence according to Definition 1 and let ccheck be a consistency function according to Definition 2. Let us suppose that the ϵ sequence is inconsistent, then $ccheck(\epsilon)$ returns a pair $< i, err >$ with $i > 0$ and an error-code $err > 0$. A Cleansing event sequence ϵ_c is a non empty sequence of events $\epsilon_c = c_1, \ldots, c_m$ able to cleanse the inconsistency identified by the error code err, namely the new event sequence $\epsilon' = e_1, \ldots, e_i, c_1, \ldots, c_m, e_{i+1}$ is consistent, i.e., $second\{ccheck(\epsilon')\} = 0$.*

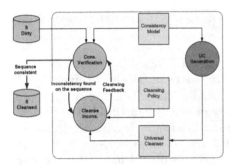
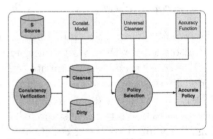

(a) The Universal Cleansing Frameworks, as presented by [32]. The UC is generated from the consistency model, then a policy is specified so that both the policy and the UC can be used for cleansing the data.

(b) The process for identifying an accurate cleansing policy. The data sequences considered consistent are fed to the policy selection process that identifies the optimal policy according to the specified accuracy function.

Fig. 2. The accurate universal cleansing framework.

Then, the *Universal Cleanser UC* can be seen as a collection that, for each error-code *err* returns the set C of the synthesised cleansing action sequences $\epsilon_c \in C$. According to the author of [34], the UC can be synthesised and used for cleansing a dirty dataset as shown in Fig. 2(a).

4   Automatic Policy Identification

The Universal Cleansing framework proposed by [34] has been considered within this work since it presents some relevant characteristics useful for our purposes, namely:

1. it is computed off-line only once on the consistency model;
2. it is *data-independent* since, once the UC has been synthesised, it can be used to cleanse *any* dataset conforming to the consistency model;
3. it is *policy-dependent* since the cleansed results may vary as the policy varies.

The Universal Cleansing framework requires a policy to drive the data cleansing activities, a *policy* (i.e., a criterion) is required for identifying which cleansing interventions has to be applied when several ones are available for a given inconsistency. The policy identification task depends on the analysis purposes and the dataset characteristics as well. This section describe how to automatically identify an optimal cleansing policy according to a (optimality) criterion specified by the user. This concept can be formalised according to the proposed framework as follows.

Definition 4 (Optimal Cleansing Policy). *Let $\epsilon = e_1, \ldots, e_n$ be an event sequence according to Definition 1 and let ccheck be a consistency function according to Definition 2.*

Let $err \in \mathbb{N}^+$ be an error-code identified on ϵ by the ccheck function and let C the set of all the cleansing action sequences $c_i \in C$ able to cleanse the error code err. A cleansing policy $\mathcal{P} : \mathbb{N} \rightarrow 2^C$ maps the error code err to a list of cleansing action sequences $c_i \in C$.

Finally, let $d : \mathbb{N}^+ \times C \times C \rightarrow \mathbb{R}$ be the distance function, an optimal cleansing policy $\mathcal{P}^*$ with respect to d assigns to each error-code the most accurate cleansing action sequence c^*, namely $\forall err \in UC, \forall c_i \in C, \exists c^{accurate}$ such that $d(err, c^*, c^{accurate}) \leq d(err, c_i, c^{accurate})$.

In Definition 4 the accuracy function basically evaluates the distance between a proposed corrective sequence (for a given error-code) against the corrective actions considered as accurate (i.e., the corrective sequence that transforms a wrong data to its real value). To identify the most accurate cleansing action, the real data value is required. Unfortunately, the real value is hard to be accessed, as discussed above, therefore a proxy is frequently used (i.e., a value considered as consistent according to the domain rules).

The source dataset is partitioned into consistent and inconsistent subsets respectively. Then, the consistent event sequences are used for identifying an accurate policy so as to cleanse the inconsistent subset. The process for accomplishing this is shortly summarized in the next few lines, the interested reader can find a more detailed description in [35].

The source dataset S is partitioned using *ccheck* into S^{tidy} and S^{dirty} which represent the consistent and inconsistent subsets of S respectively; from an event sequence ϵ (composed by $n = |\epsilon|$ events) n soiled version of ϵ are computed by removing each time one of the n events. Each soiled version ϵ'_i is cleansed in this way: the error-code err is identified (if any) and all the cleansing actions (for the specific error-code) are applied. The cleansing results are evaluated using a distance metric, the goal is to identify which cleansing action (among the available ones) can bring back ϵ'_i to ϵ. After all event sequences have been soiled and repaired, statistics are computed to identify, for each error code, which is the cleansing action (c_i hereafter) that scores the highest results i.e., the c_i that can fix the sequence most frequently. Then, the cleansing actions c_i is added to the map $\mathcal{P}$ indexed by err. At the end of the process, the map $\mathcal{P}$ is the policy.

Finally, for the sake of clarity, a graphical representation of this process has been given in Fig. 2(b).

5   Missing Fields Identification

This section will show how a machine learning approach can be used to improve data cleansing activities.

Let $\epsilon = e_1, \ldots, e_n$ be a sequence of events according to Definition 1. Referring to Definition 1, we recall that a (cleansing) *event* $e = (r_1, \ldots, r_m)$ is a *record* of the projection $(R_1, \ldots, R_m)$ over $\mathcal{Q} \subseteq \mathcal{R}$ with $m \leq l$ s.t. $r_1 \in R_1, \ldots, r_m \in R_m$. By abuse of notation we will refer to $r_1, \ldots, r_m$ as the event e fields. The event field set $ER = r_1, \ldots, r_m$ can be partitioned into 3 subsets ER_{fm}, ER_t, ER_{nc} as described next:

- ER_{fm} is the set of fields whose value can be precisely identified by the cleansing stack described in the previous sections;
- ER_t is the set of fields whose values can be trivially identified once the number of events to fix the inconsistency and the ER_{fm} values are identified;
- ER_{nc} is the set of fields whose values cannot be identified using the consistency model and the domain rules.

Considering the dataset and the inconsistency described in Fig. 1, both the *Event Type* and the *Event #* belong to ER_{fm} since the cleansing stack described in the previous sections (i.e. a cleanser built by using the Universal Cleanser and a policy for selecting the cleansing action) identifies the number of events to add (e.g., one, two, or more events), their position in the sequence (i.e. the *Event #*), and for each event the *Event Type* to be used. The *employer-id* $\in ER_t$ e.g., if a Full Time Job is ongoing with employer x and a "Full Time Cessation" *Event Type* should be added, then the event *Employer-ID* value is x. The algorithm for identifying the *Employer-ID* value is trivial (in this and in different contexts) and therefore is omitted.

The (Event) *Date* belongs to ER_{nc} since no clue about its precise value is available from the consistency model or in general from the domain rules. E.g., considering the inconsistency described in Fig. 1, the missing "Full Time Cessation" should be added between event #03 and event #04, therefore the missing event should have a date within the surrounding events dates, but no more hints can be derived from the domain rules and the consistency model to identify the precise date value.

We used a machine learning approach [3,21] to build a predictor for the ER_{nc} field value(s). The history of consistent events and the fine grained information that formal methods can provide about the event sequence evolution can be used to train a predictor for the ER_{nc} field value(s). The rationale is that the ER_{nc} field values can often be described by a statistical distribution which can be learnt using the machine learning approach. The learning is performed using consistent event sequences, once the learning phase is finished, the algorithm can be used to estimate the ER_{nc} field values for events to be added to fix inconsistent sequences.

The features candidates for the learning phase can be the following ones:

- the ER_{fm} fields. This are the main candidate features for the machine learning process.
- the ER_t fields. The ER_t fields can be omitted (totally or partially) from the feature set, since they can be obtained from the other record values, they provide little added value.
- the consistency model state and variable values. Considering the career described in Table 1, the model described in Fig. 1 can be used to enrich each event of the working history with features like the automaton state and the state variable values.
- any value derived from the previous ones.

According to the machine learning terminology, the features are split into 2 subset: X and Y whereas the Y is the target feature set i.e., the features set to be predicted once the learning phase is terminated. In our context, ER_{nc} are the Y set, while the remaining ones are the X set. Feature identification and selection is a paramount task in machine learning and it is performed (strongly) relying on domain knowledge. This topic is not further addressed. The interested reader can refer to [4,35].

A classifier or a regression function should be used for estimating the Y value. For the sake of simplicity we assume hereon that Y is composed by a single target feature, the generalization to several features is trivial. If the target feature domain is discrete a classifier is required, while if the target feature domain is continuous a regression function should be used.

The feature selection, the training activities, and the outcomes of the activities performed on the labour data are outlined in Sect. 6.4. The labour dataset is presented in Sect. 6.1.

6   Experimental Results

This section outlines how the cleansing framework described in this paper was used upon the on-line benchmark domain provided by [34].

6.1   The Labour Market Dataset

The scenario we are presenting focuses on the Italian labour market domain. According to the Italian law, every time an employer hires or dismisses an employee, or an employment contract is modified (e.g. from part-time to full-time), a *Compulsory Communication* - an event - is sent to a job registry. Each mandatory communication is stored into a record which can be decomposed into the following attributes: **e_id** and **w_id** are ids identifying the communication and the person involved respectively; **e_date** is the event occurrence date whilst **e_type** describes the event type occurring to the worker's career. The event types can be the *start*, *cessation* and *extension* of a working contract, and the *conversion* from a contract type to a different one; **c_flag** states whether the event is related to a full-time or a part-time contract while **c_type** describes the contract type with respect to the Italian law. Here we consider the *limited* (fixed-term) and *unlimited* (unlimited-term) contracts. Finally, **empr_id** uniquely identifies the employer involved.

A *communication* represents an event, whilst a career is a longitudinal data sequence whose consistency has to be evaluated. To this end, the *consistency* semantics has been derived from the Italian labour law and from the domain knowledge as follows.

c1: an employee cannot have further contracts if a full-time is active;

c2: an employee cannot have more than K part-time contracts (signed by different employers), in our context we shall assume $K = 2$;

c3: an *unlimited term* contract cannot be extended;

c4: a contract extension can change neither the contract type (*c_type*) nor the modality (*c_flag*), for instance a part-time and fixed-term contract cannot be turned into a full-time contract by an extension;

c5: a conversion requires either the *c_type* or the *c_flag* to be changed (or both).

6.2 Accurate Policy Generation Results

The process depicted in Fig. 2(b) was realised as described in Sect. 4 by using the following as input.

The Consistency Model was specified using the UPMurphi language and according to the consistency constraints specified in Sect. 6.1;

The On-line Repository provided by [34] and containing: (i) an XML dataset composed of $1,248,814$ records describing the career of $214,429$ people. Such events have been observed between 1st January 2000 and 31st December 2010; and (ii) the Universal Cleanser of the Labour Market Domain collecting all the cleansing alternatives for 342 different error-codes.

An Accuracy Function usually intended as the distance between a value v and a value v' which is considered correct or true. As the consistency we model is defined over a *set* of data items rather than a single attribute, the accuracy function should deal with all the event attribute values, as we clarified in the motivating example. To this end, the *edit distance* over the string obtained by merging the event values was used as Accuracy Function.

The accurate policy identification required about 2 h running on a x64 Linux-machine, equipped with 6 GB of RAM and connected to a MySQL DMBS for retrieving the data. The *ccheck* function basically calls the UPMurphi [43,44] planner searching for an inconsistency, the accurate policy selection as described in Sect. 4 has been implemented through C++.

The output of such a process is an *accurate policy* according to the Definition 4. Here only a summary of the results are visualized in Fig. 3. The interested reader can refer to [35] for more details on the policy produced for the labour marked dataset.

6.3 Accuracy Estimation

Here, the *k-fold cross validation* technique (see, e.g. [28]) was used to evaluate the policy synthesised according to the framework proposed.

Generally speaking, the dataset S is split into k mutually exclusive subsets (the folds) $S_1, \ldots, S_k$ of approximatively equal size. Then, k evaluation steps are performed as follows: at each step t the set of events $S \backslash S_t$ is used to identify the cleansing policy $\mathcal{P}_t$, whilst the left out subset S_t is used for evaluating the cleansing accuracy achieved by $\mathcal{P}_t$. In each step, it is computed the ratio of the event sequences the cleanser has made equal to the original via $\mathcal{P}_t$ over the cardinality of all the cleansed event sequences, namely $|S_t|$. Then, an average of

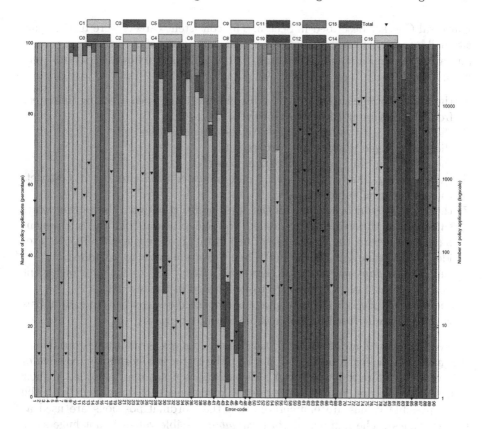

Fig. 3. A plot showing the cleansing interventions successfully performed for each error code. The x axis reports the 90 error-codes identified on the soiled dataset. The rightmost y-axis reports how many times an error-code was encountered during the policy generation process using a logarithmic scale (marked by the black triangle) while the coloured bars refer to the leftmost axis showing the percentage of cases that were successfully corrected using a specific cleansing action (Color figure online).

the k ratios is computed. More details on the *k-fold cross validation* performed on the same dataset can be found in [35].

Indeed, although $k = 10$ is usually considered as a good value in the common practice, a variable value of k ranging in $\{5, 10, 50, 100\}$ was used so that different values of accuracy $\mathcal{M}^k$ can be analysed, as summarised in Table 2. The results confirm the high-degree of accuracy reached by the policy $\mathcal{P}$ synthesised in Sect. 6.2.

6.4 Date Prediction Evaluation

This section outlines how the ER_{nc} fields (e.g., the date values) were obtained according to the process described in Sect. 5 on the Labour Market Dataset. The dataset was partitioned into inconsistent and consistent careers using the

Universal Checker described in Sect. 3. The consistent careers were further partitioned into two subsets: a training set containing 70 % of the consistent careers and a test set containing the remaining 30 % ones. The following features were used:

1. the *State* that the automaton described in Fig. 1 has when checking the career from the beginning to the current event.
2. the *statevariable* values, similarly to the previous item.
3. the *Event Type* values e.g., start, cessation, . . .
4. the *c_flag* e.g., full-time or part-time contract.
5. the *c_type* e.g., limited (fixed-term) and unlimited (unlimited-term) contract.
6. the *employer_id* value.
7. the *time_slot* which accounts for the days occurring between the known events surrounding the event to be added i.e., the previous and the subsequent ones. Considering the dataset described in Fig. 1, the event #3 and #4 dates identify the time slot for the Full Time Cessation event to be added.
8. the *elapsed_days* accounts for the days between the known previous event and the event date (i.e. the date to be guessed in the test set).
9. the *%elapsed_days* is the percentage of *elapsed_days* over the *time_slot*. This is the Y actually used in the machine learning process.

Features 1 to 7 are used as X, the *%elapsed_days* is the target variable of the machine learning process (i.e. the Y). The latter is a normalized version of feature 8, therefore it is used as Y. Features 1 to 6 are non numeric, so their values are converted into numbers using an enumeration approach (e.g., the *State* possible values are enumerated and their ordinal positions are used as feature values). The number of *state variable* possible values is not huge since the variables were extracted by the *ccheck* and the model checker using several state-variables reduction techniques (e.g. symbolic data, symmetry reduction). The interested reader can refer to [33, 34, 36] for further details.

The target feature Y has a continuous domain, therefore a regression function should be used for estimating the missing values.

Table 3 shows the regression functions results evaluation performed on the training/test dataset previously introduced using the Scikit-learn framework [42]. Two regression functions were evaluated: Linear Regression and Support Vector Regression (SVR). A polinomial combinations of the features was evaluated for every regression function, the features were combined with degree less than or equal to the specified degree.

The regression function evaluation was performed on the previously described test set. The mean squared error (between the predicted and the real "% of elapsed days" as available from the test set) ranges from 0.10 to 0.14 according to the results of Table 3. A mean squared error of 0.1 means that the estimated dates averagely differs from the real ones of about 30 % of the time slots ($30\%^2 \sim 0.1$) where the events can be placed. The results are quite interesting and could be further improved by considering some more features or by removing some outliers (a lot of careers have events ranging on time slots of several years, the errors in predicting such event date can be very huge in both absolute and

Table 2. Values of $\mathcal{M}^k$ applying k-cross validation of the policy generated.

k	5	10	50	100
$\mathcal{M}^k$	99.77%	99.78%	99.78%	99.78%
$\min(\mathcal{M}^k)$	99.75%	99.75%	99.48%	99.48%
$\max(\mathcal{M}^k)$	99.77%	99.80%	99.92%	100%
$|S_t|$	42,384	21,193	4,238	2,119
$|S|$	211,904			

Table 3. Evaluation of several regression function estimating the missing dates.

Regression function	Features polinomial degree	Mean squared error
Linear regression	1	0.11
Linear regression	2	0.10
Linear regression	3	0.10
Linear regression	4	0.13
SVR (RBF kernel)	1	0.18
SVR (RBF kernel)	2	0.14
SVR (RBF kernel)	3	0.14
SVR (RBF kernel)	4	0.14

percentual values). This topic is not further addressed since it goes beyond the scope of this paper. It is worth to highlight that the original dataset has a lot of inconsistencies and the possibility to make a career consistent by adding events whit a date estimation can be paramount for further analysis on the labour market data.

7   Related Work

Data quality analysis and improvement have been the focus of a large body of research in different domains, that involve statisticians, mathematicians and computer scientists, working in close cooperation with application domain experts, each one focusing on its own perspective [1,20]. Computer scientists developed algorithms and tools to ensure data correctness by paying attention to the whole Knowledge Discovery process, from the collection or entry stage to data visualisation [7,11,24,39]. From a statistical perspective, *data imputation* (a.k.a. data editing) is performed to replace null values or, more in general, to address quality issues while preserving the collected data statistical parameters [19].

A common technique exploited by computer scientists for data cleansing is the *record linkage* (a.k.a. *object identification, record matching, merge-purge problem*), that aims at linking the data to a corresponding higher quality version and to compare them [16]. However, when alternative (and more trusted) data are not available for linking, the most adopted solution (specially in the industry) is based on *Business Rules*, identified by domain experts for checking and cleansing dirty data. These rules are implemented in SQL or in other tool specific languages. The main drawback of this approach is that the design relies on the experience of domain experts; thus exploring and evaluating cleansing alternatives quickly become a very time-consuming task: each business rule has to be analysed and coded separately, then the overall solution still needs to be manually evaluated.

In the database area, many works focus on *constraint-based data repair* for identifying errors by exploiting FDs (Functional Dependencies) and their variants. Nevertheless, the usefulness of formal systems in databases has been motivated in [47] by observing that FDs are only a fragment of the first-order logic used in formal methods while [17] observed that, even though FDs allow one to detect the presence of errors, they have a limited usefulness since they fall short of guiding one in correcting the errors.

Two more relevant approaches based on FDs are *database repair* [9] and *consistent query answering* [2]. The former aims to find a *repair*, namely a database instance that satisfies integrity constraints and minimally differs from the original (maybe inconsistent) one. The latter approach tries to compute *consistent query answers* in response to a query i.e., answers that are true in every repair of the given database, but the source data is not fixed. Unfortunately, finding consistent answers to aggregate queries is a NP-complete problem already using two (or more) FDs [2,10]. To mitigate this problem, recently a number of works have exploited heuristics to find a database repair, as [12,29,49]. They seem to be very promising approaches, even though their effectiveness has not been evaluated on real-world domains.

More recently, the NADEEF [13] tool was developed in order to create a unified framework able to merge the most used cleansing solutions by both academy and industry. In our opinion, NADEEF gives an important contribution to data cleansing also providing an exhaustive overview about the most recent (and efficient) solutions for cleansing data. Indeed, as the authors remark, consistency requirements are usually defined on either a single tuple, two tuples or a set of tuples. The first two classes are enough for covering a wide spectrum of basic data quality requirements for which FD-based approaches are well-suited. However, the latter class of quality constraints (that NADEEF does not take into account according to its authors) requires reasoning with a (finite but not bounded) set of data items over time as the case of longitudinal data, and this makes the exploration-based technique a good candidate for that task.

Finally, the work of [12], to the best of our knowledge, can be considered a milestone in the identification of accurate data repairs. Basically, they propose (i) a heuristic algorithm for computing a data repair satisfying a set of constraints expressed through *conditional* functional dependencies and (ii) a method for evaluating the accuracy of the repair. Compared with our work their approach differs mainly for two aspects. First, they use conditional FDs for expressing constraints as for NADEEF, while we consider constraints expressed over more than two tuples at a time. Second, their approach for evaluating the accuracy of a repair is based upon the user feedback. Indeed, to reduce the user effort only a *sample* of the repair - rather than the entire repair - is proposed to the user according to authors. Then a confidence interval is computed for guaranteeing a precision higher than a specified threshold. On the contrary, our approach limits the user effort only to the modelling phase, exploiting a graph search for computing accurate cleansing alternatives.

8   Concluding Remarks and Further Directions

In this paper the model based cleansing framework outlined in [34, 35] was extended using a machine learning approach. The cleansing framework can produce the Universal Cleanser: a taxonomy of possible errors and for each error the set of possible corrections, with respect to a consistency model describing the date expected behaviour. A policy (selecting one correction when several ones are available for fixing an error) can be added to the framework. Several policies can be used achieving different cleansing results, an optimality criterion can be specified and the framework can identify the policy that best fit the optimality criterion with respect to the data to be analyzed. To build an effective cleanser the last step is to identify the field values for which no clue is available from the consistency model or in general from the domain rules. E.g., considering an ordered sequence of timestamped events, the consistency model can identify that an event is missing between the $k - 1^{th}$ and the kth event, but no clue is available for identifying the (missing) event timestamp. This paper enriched the cleansing framework adding a machine learning approach whereas consistent portions of the dataset are used to train a learning algorithm that can be later used to estimate the missing information. The rationale is that the missing data can often be described by a statistical distribution which can be learnt using the machine learning approach. The joint use of model based and machine learning approach provides a powerful instrument since: (1) it makes the development of cleansing routines easier, because the user can specify a triple <data consistency model, policy (or optimality criterion for identifying a policy), a data training set (or a criterion for extracting the training dataset)> and then the cleansing routines are generated automatically, thus greatly reducing the required human effort. It is worth to note that the Universal Cleanser can be used to ensure that the selected training dataset has no inconsistency; (2) it allows the data analysts to quickly experiment several cleansing options and to evaluate the cleansing results; (3) the model based cleansing ensures that the cleansed data fit the consistency model provided by the data analysts and machine learning is used to improve the outcome achieving fine grained results.

The proposed hybrid framework represents the final step to rapidly produce accurate and reliable cleansing procedures.

Finally, our approach has been successfully tested for generating an accurate policy over a real (weakly-structured) dataset describing people working careers.

As a further step, we have been working toward evaluating the effectiveness of our approach on biomedical domain.

References

1. Abello, J., Pardalos, P.M., Resende, M.G.: Handbook of Massive Data Sets, vol. 4. Springer, US (2002)
2. Bertossi, L.: Consistent query answering in databases. ACM Sigmod Rec. **35**(2), 68–76 (2006)

3. Bishop, C.M., et al.: Pattern Recognition and Machine Learning, vol. 1. Springer, New York (2006)
4. Blum, A.L., Langley, P.: Selection of relevant features and examples in machine learning. Artif. Intell. **97**(1), 245–271 (1997)
5. Boselli, R., Cesarini, M., Mercorio, F., Mezzanzanica, M.: Inconsistency knowledge discovery for longitudinal data management: a model-based approach. In: Holzinger, A., Pasi, G. (eds.) HCI-KDD 2013. LNCS, vol. 7947, pp. 183–194. Springer, Heidelberg (2013)
6. Boselli, R., Cesarini, M., Mercorio, F., Mezzanzanica, M.: Planning meets data cleansing. In: The 24th International Conference on Automated Planning and Scheduling (ICAPS), pp. 439–443. AAAI (2014)
7. Boselli, R., Cesarini, M., Mercorio, F., Mezzanzanica, M.: A policy-based cleansing and integration framework for labour and healthcare data. In: Holzinger, A., Jurisica, I. (eds.) Knowledge Discovery and Data Mining. LNCS, vol. 8401, pp. 141–168. Springer, Heidelberg (2014)
8. Boselli, R., Cesarini, M., Mercorio, F., Mezzanzanica, M.: Towards data cleansing via planning. Intelligenza Artificiale **8**(1), 57–69 (2014)
9. Chomicki, J., Marcinkowski, J.: Minimal-change integrity maintenance using tuple deletions. Inf. Comput. **197**(1), 90–121 (2005)
10. Chomicki, J., Marcinkowski, J.: On the computational complexity of minimal-change integrity maintenance in relational databases. In: Bertossi, L., Hunter, A., Schaub, T. (eds.) Inconsistency Tolerance. LNCS, vol. 3300, pp. 119–150. Springer, Heidelberg (2005)
11. Clemente, P., Kaba, B., Rouzaud-Cornabas, J., Alexandre, M., Aujay, G.: SPTrack: visual analysis of information flows within SELinux policies and attack logs. In: Huang, R., Ghorbani, A.A., Pasi, G., Yamaguchi, T., Yen, N.Y., Jin, B. (eds.) AMT 2012. LNCS, vol. 7669, pp. 596–605. Springer, Heidelberg (2012)
12. Cong, G., Fan, W., Geerts, F., Jia, X., Ma, S.: Improving data quality: consistency and accuracy. In: Proceedings of the 33rd International Conference on Very Large Data Bases, pp. 315–326. VLDB Endowment (2007)
13. Dallachiesa, M., Ebaid, A., Eldawy, A., Elmagarmid, A.K., Ilyas, I.F., Ouzzani, M., Tang, N.: Nadeef: a commodity data cleaning system. In: Ross, K.A., Srivastava, D., Papadias, D. (eds.) SIGMOD Conference, pp. 541–552. ACM (2013)
14. De Silva, V., Carlsson, G.: Topological estimation using witness complexes. In: Proceedings of the First Eurographics Conference on Point-Based Graphics, pp. 157–166. Eurographics Association (2004)
15. Devaraj, S., Kohli, R.: Information technology payoff in the health-care industry: a longitudinal study. J. Manag. Inf. Syst. **16**(4), 41–68 (2000)
16. Elmagarmid, A.K., Ipeirotis, P.G., Verykios, V.S.: Duplicate record detection: a survey. IEEE Trans. Knowl. Data Eng. **19**(1), 1–16 (2007)
17. Fan, W., Li, J., Ma, S., Tang, N., Yu, W.: Towards certain fixes with editing rules and master data. In: Proceedings of the VLDB Endowment, vol. 3(1–2), pp. 173–184 (2010)
18. Fayyad, U., Piatetsky-Shapiro, G., Smyth, P.: The kdd process for extracting useful knowledge from volumes of data. Commun. ACM **39**(11), 27–34 (1996)
19. Fellegi, I.P., Holt, D.: A systematic approach to automatic edit and imputation. J. Am. Stat. Assoc. **71**(353), 17–35 (1976)
20. Fisher, C., Lauría, E., Chengalur-Smith, S., Wang, R.: Introduction to Information Quality. AuthorHouse, USA (2012)
21. Freitag, D.: Machine learning for information extraction in informal domains. Mach. Learn. **39**(2–3), 169–202 (2000)

22. Hansen, P., Järvelin, K.: Collaborative information retrieval in an information-intensive domain. Inf. Process. Manag. **41**(5), 1101–1119 (2005)
23. Holzinger, A.: On knowledge discovery and interactive intelligent visualization of biomedical data - challenges in human-computer interaction & biomedical informatics. In: Helfert, M., Francalanci, C., Filipe, J. (eds.) DATA. SciTePress (2012)
24. Holzinger, A., Bruschi, M., Eder, W.: On interactive data visualization of physiological low-cost-sensor data with focus on mental stress. In: Cuzzocrea, A., Kittl, C., Simos, D.E., Weippl, E., Xu, L. (eds.) CD-ARES 2013. LNCS, vol. 8127, pp. 469–480. Springer, Heidelberg (2013)
25. Holzinger, A., Yildirim, P., Geier, M., Simonic, K.M.: Quality-based knowledge discovery from medical text on the web. In: Pasi et al. [38], pp. 145–158
26. Holzinger, A., Zupan, M.: Knodwat: a scientific framework application for testing knowledge discovery methods for the biomedical domain. BMC Bioinf. **14**, 191 (2013)
27. Kapovich, I., Myasnikov, A., Schupp, P., Shpilrain, V.: Generic-case complexity, decision problems in group theory, and random walks. J. Algebra **264**(2), 665–694 (2003)
28. Kohavi, R.: A study of cross-validation and bootstrap for accuracy estimation and model selection. In: Proceedings of the 14th International Joint Conference on Artificial Intelligence, IJCAI 1995, vol. 2, pp. 1137–1143. Morgan Kaufmann Publishers Inc., San Francisco (1995). http://dl.acm.org/citation.cfm?id=1643031.1643047
29. Kolahi, S., Lakshmanan, L.V.: On approximating optimum repairs for functional dependency violations. In: Proceedings of the 12th International Conference on Database Theory, pp. 53–62. ACM (2009)
30. Lovaglio, P.G., Mezzanzanica, M.: Classification of longitudinal career paths. Qual. Quant. **47**(2), 989–1008 (2013)
31. Madnick, S.E., Wang, R.Y., Lee, Y.W., Zhu, H.: Overview and framework for data and information quality research. J. Data Inf. Qual. **1**(1), 2:1–2:22 (2009)
32. Mezzanzanica, M., Boselli, R., Cesarini, M., Mercorio, F.: Data Quality through Model Checking Techniques. In: Gama, J., Bradley, E., Hollmén, J. (eds.) IDA 2011. LNCS, vol. 7014, pp. 270–281. Springer, Heidelberg (2011)
33. Mezzanzanica, M., Boselli, R., Cesarini, M., Mercorio, F.: Data quality sensitivity analysis on aggregate indicators. In: Helfert, M., Francalanci , C., Filipe, J. (eds.) DATA 2012-The International Conference on Data Technologies and Applications, pp. 97-108. SciTePress (2012). 10.5220/0004040300970108
34. Mezzanzanica, M., Boselli, R., Cesarini, M., Mercorio, F.: Automatic synthesis of data cleansing activities. In: Helfert, M., Francalanci, C. (eds.) The 2nd International Conference on Data Management Technologies and Applications (DATA), pp. 138–149. Scitepress (2013)
35. Mezzanzanica, M., Boselli, R., Cesarini, M., Mercorio, F.: Improving data cleansing accuracy: a model-based approach. In: The 3rd International Conference on Data Technologies and Applications, pp. 189–201. Insticc (2014)
36. Mezzanzanica, M., Boselli, R., Cesarini, M., Mercorio, F.: A model-based evaluation of data quality activities in KDD. Inf. Process. Manag. **51**(2), 144–166 (2015). doi:10.1016/j.ipm.2014.07.007
37. Mezzanzanica, M., Boselli, R., Cesarini, M., Mercorio, F.: A model-based approach for developing data cleansing solutions. ACM J. Data Inf. Qual. **5**(4), 1–28 (2015). doi:10.1145/2641575

38. Ng, A.Y.: Feature selection, l 1 vs. l 2 regularization, and rotational invariance. In: Proceedings of the Twenty-first International Conference on Machine Learning, p. 78. ACM (2004)

39. de Oliveira, M.C.F., Levkowitz, H.: From visual data exploration to visual data mining: a survey. IEEE Trans. Vis. Comput. Graph. 9(3), 378–394 (2003)

40. Pasi, G., Bordogna, G., Jain, L.C.: An introduction to quality issues in the management of web information. In: Quality Issues in the Management of Web Information [38], pp. 1–3

41. Pasi, G., Bordogna, G., Jain, L.C. (eds.): Quality Issues in the Management of Web Information. Intelligent Systems Reference Library, vol. 50. Springer, Heidelberg (2013)

42. Pedregosa, F., Varoquaux, G., Gramfort, A., Michel, V., Thirion, B., Grisel, O., Blondel, M., Prettenhofer, P., Weiss, R., Dubourg, V., Vanderplas, J., Passos, A., Cournapeau, D., Brucher, M., Perrot, M., Duchesnay, E.: Scikit-learn: machine learning in Python. J. Mach. Learn. Res. 12, 2825–2830 (2011)

43. Penna, G.D., Intrigila, B., Magazzeni, D., Mercorio, F.: UPMurphi: a tool for universal planning on pddl+ problems. In: Proceedings of the 19th International Conference on Automated Planning and Scheduling (ICAPS 2009), pp. 106–113. AAAI Press, Thessaloniki, Greece (2009). http://aaai.org/ocs/index.php/ICAPS/ICAPS09/paper/view/707

44. Penna, G.D., Magazzeni, D., Mercorio, F.: A universal planning system for hybrid domains. Appl. Intell. 36(4), 932–959 (2012). doi:10.1007/s10489-011-0306-z

45. Prinzie, A., Van den Poel, D.: odeling complex longitudinal consumer behavior with dynamic bayesian networks: an acquisition pattern analysis application. J. Intell. Inf. Syst. 36(3), 283–304 (2011)

46. Rahm, E., Do, H.: Data cleaning: problems and current approaches. IEEE Data Eng. Bull. 23(4), 3–13 (2000)

47. Vardi, M.: Fundamentals of dependency theory. In: Borger, E. (ed.) Trends in Theoretical Computer Science, pp. 171–224. Computer Science Press, Rockville (1987)

48. Wang, R.Y., Strong, D.M.: Beyond accuracy: what data quality means to data consumers. J. Manag. Inf. Syst. 12(4), 5–33 (1996)

49. Yakout, M., Berti-Équille, L., Elmagarmid, A.K.: Don't be scared: use scalable automatic repairing with maximal likelihood and bounded changes. In: Proceedings of the 2013 International Conference on Management of Data, pp. 553–564. ACM (2013)

Social Influence and Influencers Analysis: A Visual Perspective

Chiara Francalanci and Ajaz Hussain[✉]

Department of Electronics, Information and Bio-Engineering,
Politecnico di Milano, 20133 Milan, Italy
{francala, hussain}@elet.polimi.it

Abstract. Identifying influencers is an important step towards understanding how information spreads within a network. Social networks follow a power-law degree distribution of nodes, with a few hub nodes and a long tail of peripheral nodes. While there exist consolidated approaches supporting the identification and characterization of hub nodes, research on the analysis of the multi-layered distribution of peripheral nodes is limited. In social media, hub nodes represent social influencers. However, the literature provides evidence of the multi-layered structure of influence networks, emphasizing the distinction between influencers and influence. Information seems to spread following multi-hop paths across nodes in peripheral network layers. This paper proposes a visual approach to the graphical representation and exploration of peripheral layers and clusters by exploiting the theory of k-shell decomposition analysis. The core concept of the proposed approach is to partition the node set of a graph into pre-defined hub and peripheral nodes. Then, a power-law based modified force-directed method is applied to clearly display local multi-layered neighborhood clusters around hub nodes based on a characterization of the content of message that we refer to as *content specificity*. We put forward three hypotheses that allow the graphical identification of the peripheral nodes that are more likely to be influential and contribute to the spread of information. Hypotheses are tested on a large sample of tweets from the tourism domain.

Keywords: Semantic networks · Power law graphs · Social media influencers · Social media influence

1 Introduction

The social media literature makes a distinction between influencers and influence. Influencers are prominent social media users with a broad audience. For example, social users with a high number of followers and retweets on Twitter, or a multitude of friends on Facebook, or a broad connections on LinkedIn. The term influence refers to the social impact of the content shared by social media users. In traditional media, such as TV or radio, the coverage of the audience was considered the primary indicator of influence. However, traditional media are based on broadcasting rather than communication, as opposed to social media, which truly interactive. Influencers may say something totally uninteresting and, as a consequence, they obtain little or no attention. On the contrary, if social media users seem to be interested in something, they normally

© Springer International Publishing Switzerland 2015
M. Helfert et al. (Eds.): DATA 2014, CCIS 178, pp. 81–98, 2015.
DOI: 10.1007/978-3-319-25936-9_6

show it by participating in the conversation with a variety of mechanisms, mostly by sharing the content that they have liked. [2]; [25] has noted that a content that has an impact on a user's mind is usually shared. Influencers are prominent social media users, but we cannot be certain that the their shared content has influence, as discussed by [7].

In previous research, [20] also have shown how the content of messages can play a vital role. Regardless of the author's centrality, content can be a determinant of the author's social influence. Results suggest that peripheral nodes can be influential. This paper starts from the observation made by [12] social networks of influence follow a power-law distribution function, with a few hub nodes and a long tail of peripheral nodes. In social media, hub nodes represent social influencers [28], but influential content can be generated by peripheral nodes and spread along possibly multi-hop paths originated in peripheral network layers. The ultimate goal of our research is to understand how influential content spreads across the network. For this purpose, identifying and positioning hub nodes is not sufficient, while we need an approach that supports the exploration of peripheral nodes and of their mutual connections. In this paper, we exploit a modified power-law based force-directed algorithm [17]. The algorithm is based on the idea that hub nodes should be prioritized in laying out the overall network topology, but their placement should depend on the topology of peripheral nodes around them. In our approach, the topology of the periphery is defined by grouping peripheral nodes based on the strength of their link to hub nodes, as well as the strength of their mutual interconnections, which is the metaphor of k-shell decomposition analysis [19].

The approach is tested on a large sample of tweets expressing opinions on a selection of Italian locations relevant for tourism. Tweets have been semantically processed and tagged with information on (a) the location to which they refer, (b) the location's brand driver (or category) on which authors express an opinion, (c) the number of retweets, and (d) the identifier of the retweeting author. With this information, we draw corresponding multi-mode networks highlighting the connections among authors (retweeting) and their interests (brand or category). By visually exploring and understanding the multi-layered periphery of nodes, we also propose a few content-related hypotheses in order to understand the relationship among *frequency*, *specificity*, and *retweets* of posts. Results highlight the effectiveness of our approach, providing interesting visual insights on how unveiling the structure of the periphery of the network can visually show the role of peripheral nodes in determining influence. Empirical results show that peripheral nodes also plays a vital role and can be a determinant of social influence. The main innovative aspect of our approach is that we use statistics (hypotheses) and visualization together. One can visually verify the proposed hypotheses on graphs.

The remainder of this paper is structured as follows. Section 2 discusses limitations of existing network drawing techniques, and provides insights about influence and influencers in social media. Section 3 presents the proposed research hypotheses. Section 4 discusses the implementation aspects. Section 5 presents the experimental methodology, performance evaluation, and empirical results. Conclusions are drawn in Sect. 6.

2 State of the Art

In this section, we will discuss about limitations of existing network visualization techniques. We will also explore the concept of influencers and influence in social media.

2.1 Network Visualization Techniques

Several research efforts in network visualization have targeted power-law algorithms and their combination with the traditional force-directed techniques, as for example in [1]. Among these approaches, the most notable is the Out-Degree Layout (ODL) for the visualization of large-scale network topologies, presented by [27]. The core concept of the algorithm is the segmentation of network nodes into multiple layers based on their out-degree, i.e. the number of outgoing edges of each node. The positioning of network nodes starts from those with the highest out-degree, under the assumption that nodes with a lower out-degree have a lower impact on visual effectiveness.

The topology of the network plays an important role such that there are plausible circumstances under which nodes with a higher number of connections or greater betweeness have little effect on the range of a given spreading process. For example, if a hub exists at the end of a branch at the periphery of a network, it will have a minimal impact in the spreading process through the core of the network, whereas a less connected person who is strategically placed in the core of the network will have a significant effect that leads to dissemination through a large fraction of the population. To identify the core and the multi-layered periphery of the clustered network, we use a technique based on the metaphor of k-shell (also called k-core) decomposition of the network, as discussed in [19].

2.2 Influencers and Influence in Social Networks

Traditionally, the literature characterizes a social media user as an influencer on the basis of structural properties. Centrality metrics are the most widely considered parameters for the structural evaluation of a user's social network. The centrality of a concept has been defined as the significance of an individual within a network [14]. Centrality has attracted a considerable attention as it clearly recalls concepts like social power, influence, and reputation. A node that is directly connected to a high number of other nodes is obviously central to the network and likely to play an important role [6]. Freeman [16] introduced the first centrality metrics, named as *degree centrality*, which is defined as the number of links incident upon a node. A node with many connections to other nodes, likely to play an important role [29]. A distinction is made between in-degree and out-degree centrality, measuring the number of incoming and outgoing connections respectively. This distinction has also been considered important in social networks. For example, Twitter makes a distinction between friends and followers. Normally, on Twitter, users with a high in-degree centrality (i.e. with a high number of followers) are considered influencers. In addition to degree centrality, the literature also shows other structural metrics for the identification of influencers in social networks. Leavitt [22] presented an approach, where users were identified as influencers based on their total number of retweets. Results highlighted how the number of retweets are positively correlated with the level of users' activity (number of tweets) and their in-degree centrality (number of followers). Besides structural metrics, the more recent literature has associated the complexity of the concept of influence with the variety of content. Several research works have addressed the need for considering content-based metrics of influence [8]. Content metrics such as the number of mentions, URLs, or hashtags have been proved to increase the probability of retweeting [5].

While the literature provides consolidated approaches supporting the identification and characterization of hub nodes i.e. influencers in a social network, research on information spread, which is multi-layered distribution of peripheral nodes, is limited. The literature mainly focuses on the concept of influencers, while there is a need for effective visualization techniques in social networks, which enable users to visually explore large-scale complex social networks to identify the users who are responsible for influence. This paper presents a power-law based modified force-directed technique, that extends a previous algorithm discussed in [17] by exploiting the k-shell decomposition technique [19]. The algorithm is briefly summarized in Sect. 4.

3   The Power-Law Algorithm

This section provides a high-level description of the graph layout algorithm used in this paper. An early version of the algorithm has been presented by [15, 17]. This paper improves the initial algorithm by identifying multiple layers of peripheral nodes around hub nodes according to the k-shell decomposition approach. The power-law layout algorithm belongs to the class of force-directed algorithms, such as the one by [12]. In this algorithm, the `NodePartition()` step is a pre-processing method aimed at distinguishing hub nodes from peripheral nodes. This step is performed by pre-identifying hub nodes as N_h, which represents one of the following two sets:

1. A set of predefined tourism destinations, called *brands*, i.e. *Amalfi, Amalfi Coast, Lecce, Lucca, Naples, Palermo* and *Rome* (7 in total).
2. A set of predefined brand drivers of a destination's brand, called *categories*. Examples of categories are *Art & Culture, Food & Drinks, Events & Sport, Services & Transports*, etc., as explained in Sect. 5.

The following code snippet provides a high-level overview of the whole algorithm by showing its main building blocks. The proposed approach is aimed at the exploitation of the power-law degree distribution of author nodes (N_p). Provided that the distribution of the degree of the nodes follows a power law, we partition the network into bipartite graph of two disjoint vertices N into the set of predefined hub nodes N_h, which represents topics (brands or categories), and the set of peripheral nodes N_p, which represents authors, such that $N = N_h \cup N_p$, with $N_h \cap N_p = \emptyset$.

DATA:

N_h = Hub Nodes representing Brands or Categories;

N_p = Peripheral Nodes representing authors;

E = Edges connecting authors to either brands or categories whenever one of author's tweet refers to that brand or category;

d = Degree of author node representing the number of edges connected to author node;

t_p = Number of times the author N_p tweeted about N_h (i.e. Brand or Category);

T = Energy/Temperature Variable; T_h = Temperature threshold, to control simulation.

BEGIN

```
1. NodePartition();
2. InitialLayout();
   IF (T > T_h) DO
       AttractionForce(N_h,N_p);
       RepulsionForce(N_h,E);
   ELSE
       AttractionForce(N_p,N_h);
       RepulsionForce(N_p,E);
3. LShellDecomposition(N_p,t_p);
4. NodesPlacement (N_h,N_p,t_p);
5. TempCoolDown(T);
6. resetNodesSizes(N_p,N_t,d);
```

END

The InitialLayout() step provides the calculation of attraction and repulsion forces, based upon value of T_h, which is a threshold value that can be tuned to optimize the layout, by providing maximum forces exerted upon hub-nodes N_h. The formulae of attraction and repulsion forces are similar to those used in traditional force-directed approaches, such as [12]. In this paper, the forces formulae have been taken from the power-law based modified force-directed algorithm as presented in [17].

The LShellDecomposition(N_p, t_p) method is responsible for the calculation of the l-shell value of peripheral nodes in N_p. We tuned this technique by means of the metaphor of k-shell decomposition analysis [11], in order to define the concept of *level* of each node in the multi-layered periphery of our graphs. This process assigns an integer as level index (l_s) to each node, representing its location according to successive layers (l shells) in the network. In this way, the author nodes who tweeted once about a specific topic, will have ($l_s = 1$) forming the outmost layer around that topic, and those who tweeted twice will have ($l_s = 2$) forming the inward successive layer, and so on. By this metaphor, small values of (l_s) define the periphery of the network (outliers), while the innermost network levels correspond to greater values of l_s, containing those authors who tweeted most frequently, as shown in Fig. 1.

The NodePlacement(N_h, N_p, T_p) step performs the placement of nodes on graph canvas based on the computation of forces among nodes. The placement of N_p nodes will be as per their T_p value in order to create multi-layered peripheral layers of author nodes N_p around hub-nodes N_h. The `TempCooldown(T)` step is responsible for the control of the overall iteration mechanism. This step is responsible for cooling down the system temperature, in order to make the algorithm converge. We introduce a customized dynamic temperature cool down scheme, which adapts the iterative step based on the current value of temperature. The temperature is supposed to be initialized at a value T_{start}, and then to be reduced by a variable Δt based on the current value of the temperature itself. This approach provides a convenient way to adapt the speed of iteration of the algorithm to the number of nodes to be processed. While processing hub nodes (a few), the temperature decreases slowly; while processing peripheral nodes (many), the temperature decreases more rapidly to avoid expensive computations for nodes that are not *central* to the overall graph layout. Finally, the `resetNodesSizes(N_p,N_t,d)` method is responsible for resetting the sizes of each node in the graph, based upon their degree. The higher the degree of a node, the greater the size and vice versa.

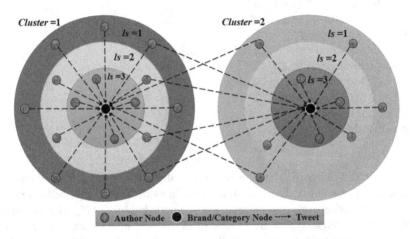

Fig. 1. Metaphor of k-shell decomposition analysis.

4   Research Hypotheses

The literature indicates that social media are associated with a long-tail effect, with a variety of smaller communities [23]. While general content has a broad audience, there exists a variety of smaller communities who are interested in specific content. Such long-tail effect suggests that these communities are numerous and their specific interests are virtually limitless [14]. Social media users also consider *specificity* as an important metric for making behavioral decisions [10]. The *specificity* of shared content by social media users can be described as the level of detail with which a user comments on a given subject of interest. Klotz [20] has shown how the content of messages can play a critical role and can be a determinant of the social influence of a message irrespective of the centrality of the message's author. Twitter users with a high volume of tweets can be referred to as 'information sources' or 'generators' [18]. The literature also shows that social media user intend to post content that is shared frequently by many other users [4]. Social media users wish to be influential [13]. Intuitively, since users want to be interesting to many, if a user talks a lot, he/she will probably address the needs of multiple specific communities, i.e. multiple topics. Consequently, our first hypothesis posits a positive association between frequency of tweets and content specificity in multiple topics.

- **H1:** Tweeting with a high frequency of tweets is positively associated with the number of topics (brands or categories) (i.e. visually, *potential influencers* are the peripheral authors).

If a speaker builds an audience around specific shared interests, content specificity may have a positive, as opposed to negative impact on audience attention. The literature suggests that social media user intend to post content that shared frequently by many other users [4]. The literature also explains that retweeting is associated with information sharing, commenting or agreeing on other peoples' messages and entertaining followers [9]. Kwak [21] also show that the most trending topics have an active period of one week, while half of retweets of a given tweet occurs within one hour and 75 % within one day. The frequency of retweets is a major factor for estimating the quality of posts. It can be an important criterion since users tend to retweet valuable posts [13]. In the communities of people who are interested in specific content, users share specific content that followers are more likely to retweet. Intuitively, if a user tweets about multiple topics, interesting to many specific and active communities, he/she is most likely to get more retweets. Consequently, in the following hypothesis we posit a positive association between the number of topics and the frequency of retweets.

- **H2:** Tweeting about multiple topics (brands or categories) is positively associated with the frequency of retweets (i.e. visually, peripheral authors, connected to multiple topics, are *actual influencers*).

The breadth of the audience was considered the first and foremost indicator of influence for traditional media, such as television or radio. However, traditional media are based on broadcasting rather than communication, while social media are truly interactive [7]. In traditional media, influencers intend to target a large audience by broadcasting frequently. Similarly, in social media, e.g. in twitter, influencers intend to be more interactive by showing their presence and frequently tweeting [10]. If social media users are interested in something, they typically show it by participating in the conversation with a variety of mechanisms and, most commonly, by frequently sharing the content that they have liked [28]. A content that has had an impact on a user's mind is shared and gathers attention by others. The volumes of retweets are positively correlated with the level of users' activity (number of tweets) and their in-degree centrality (number of followers), as noted by [22]. In social media, while sharing content, users may be referred as '*generalists*' or '*information sources*' who talk about multiple topics [18]. On the contrary, there exist such users, who are very specific in sharing content related to specific topic or brand. These specific authors seems to be potential influence spreaders [14]. We posit that, these authors have to be active participants in each community by talking a lot. Our third hypothesis posits that such authors have a greater probability of being retweeted due to frequent tweets, and can be both potential and actual influencers.

- **H3:** Tweeting more frequently (with a high frequency) about a single topic (brand or category) is positively associated with the frequency of retweets (i.e. visually, authors, drawn closer to single topic, are both *actual* and *potential* influencers).

We posit the aforementioned three hypotheses that tie *content specificity*, *frequency of tweets* and *frequency of retweets*. Visually, hypothesis H1 can be verified by observing the peripheral authors positioned in the outer-most layers of each cluster (lowest 1-shell value, l_s=1), which are only connected to one cluster hub (brand or category). These authors seems to be talking about a single brand or category. Such outlier authors can be *potential* influencers, if they further connect to other authors via content sharing and tweeting about multiple topics (brands or categories). Similarly, hypothesis H2 can be visually verified by observing authors who are placed in between multiple clusters, connected to multiple clusters' hubs (brands or categories), and seem to be talking about multiple topics. These authors are *actual* influencers as they receive a high number of retweets by tweeting about multiple topics. Moreover, hypothesis H3 can be visually verified by observing those authors who are positioned in the inner-most periphery of a single cluster (highest l_s value) and seem to be placed close to the cluster hub (brand or category). Such authors are both *actual* and *potential* influencers as they are most specific about content sharing. These authors tweet frequently about a single topic (brand or category) and receive a high number of retweets.

5 Experimental Methodology and Results

In this section, we will present the dataset that we have used in our experiment and the network models that we have built from the dataset. Empirical evaluations and related visualization results are also presented in this section.

5.1 Variable Definition and Operationalization

Each graph G (A, T) has a node set A representing authors and an edge set T representing tweets. We define as $N_T (a)$ the total number of tweets posted by author a. We define as $N_R (a)$ total number of times author a, has been retweeted. Tweets can refer to a brand b or to a category c. We define as $N_B (a)$ the total number of brands mentioned by each author a, in all his/her tweets, i.e. *brand specificity*. Similarly, $N_C (a)$ represents the total number of categories mentioned by each author a, in all his/her tweets, i.e. *category specificity*.

5.2 Data Sample

We collected a sample of tweets over a two-month period (December 2012 – January 2013). For the collection of tweets, we queried the public Twitter APIs by means of an automated collection tool developed ad-hoc. Twitter APIs have been queried with the following crawling keywords, representing tourism destinations (i.e. brands): *Amalfi*, *Amalfi Coast*, *Lecce*, *Lucca*, *Naples*, *Palermo* and *Rome*. Two languages have been considered, *English* and *Italian*. Collected tweets have been first analyzed with a proprietary semantic engine [6] in order to tag each tweet with information about *a)* the location to which it refers, *b)* the location's brand driver (or category) on which authors express an opinion, *c)* the number of retweets (if any), and *d)* the identifier of the retweeting author. Our data sample is referred to the tourism domain. We have adopted a modified version of the Anholt Nation Brand index model to define a set of categories of content referring to specific brand drivers of a destination's brand [2]. Examples of brand drivers are *Art & Culture*, *Food & Drinks*, *Events & Sport*, *Services & Transports*, etc. A tweet is considered *Generic* if it does not refer to any *Specific* brand driver, while it is considered *Specific* if it refers to at least one of Anholt's brand drivers. Tweets have been categorized by using an automatic semantic text processing engine that has been developed as part of this research. The semantic engine can analyze a tweet and assign it to one or more semantic categories. The engine has been instructed to categorize according to the brand drivers of Anholt's model, by associating each brand driver with a specific content category described by means of a network of keywords. Each tweet can be assigned to multiple categories. We denote with N_C the number of categories each tweet w is assigned to; the specificity $S(w)$ of a given tweet w is defined as in Eq. 1, while Table 1 refer to the descriptive statistics of the original non-linear variables.

$$S(w) = \begin{cases} 0, N_c = 0 \\ 1, N_c > 0 \end{cases}. \tag{1}$$

Table 1. Basic descriptive statistics of our dataset.

Variable	Value	S.D.
Number of tweets	957,632	–
Number of retweeted tweets	79,691	–
Number of tweeting authors	52,175	–
Number of retweets	235,790	–
Number of retweeting authors	66,227	–
Average number of tweets per author	10.07	±86.83
Average number of retweeted tweets per author	1.525	±4.67
Average number of retweets per author	1.40	±4.52
Average frequency of retweets per author	0.58	±0.38
Average content Specificity per author	0.35	±0.46

5.3 Network Models

In order to verify the effectiveness of the proposed algorithm with respect to the goal of our research, we have defined different network models based on the data set described in the previous section. Figure 2 provides an overview of the adopted network models.

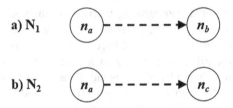

Fig. 2. Network models a) N_1: Author $\rightarrow$ Brand, b) N_2: Author $\rightarrow$ Category.

- Author $\rightarrow$ Brand (N_1). This model considers the relationship among authors and domain brands, i.e., touristic destinations in our data set. The network is modelled as bipartite graph, where an author node n_a is connected to a brand node n_b whenever author a has mentioned brand b in at least one of his/her tweets.
- Author $\rightarrow$ Category (N_2). This model considers the relationship among authors and domain brand drivers (categories), i.e., city brand drivers in our data set (namely, *Arts & Culture, Events & Sports, Fares & Tickets, Fashion & Shopping, Food & Drink, Life & Entertainment, Night & Music, Services & Transport, and Weather & Environmental*). The network is modelled as bipartite graph, where an author node n_a is connected to a category node n_c whenever author a has mentioned a subject belonging to category c in at least one of his/her tweets.

5.4   Network Visualization

The empirical results and discussions on network visualization will adopt network N_1 network (i.e. Author $\rightarrow$ Brand) as reference example. Figure 3 provides an enlarged view of network N_1 visualized by means of the proposed power-law layout algorithm. A summary description for N_1 and N_2 networks is presented in Table 2, where N_R (a) represents the total number of retweets, N_B (a) shows the total number of tweets in which author a talked about brand B (N_1 network), N_c (a) shows the total number of tweets in which author a talked about category C (N_2 network), and N_T (a) represents the total frequency of author a (i.e. the total number of tweets of author a).

The network visualization depicted in Fig. 3 adopts multicolor nodes to represent authors, and highlighted encircled blue (dark) nodes to represent tourism destinations (i.e. brands) on which authors have expressed opinions in their tweets. The layout of the network produced by the power-law layout algorithm clearly highlights that author nodes aggregate in several groups and subgroups based on their connections with brand nodes, which in this case are the hub nodes.

5.5   Empirical Results

This section reports on the empirical testing and evaluation of the proposed hypotheses. First, we discuss our research model and then we present empirical results.

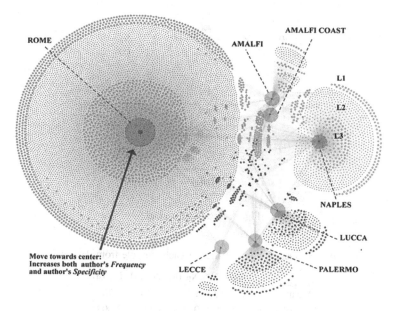

Fig. 3. Network N_1: Author $\rightarrow$ Brand (enlarged view).

Table 2. Descriptive statistics on the dimensions of N_1 and N_2 networks.

Authors	N_R (a)	N_B (a)	N_C (a)	N_T (a)
398	92	856	1,913	2,769
1,662	364	2,905	5,959	8,864
10,710	2,907	12,559	18,498	31,057
18,711	5,329	21,140	29,842	50,982
30,310	8,690	33,684	46,120	79,804
37,626	10,529	41,620	56,960	98,580
47,295	12,833	52,208	71,667	1,23,875

5.5.1 Research Model

AMOS 20 [3] has been used to analyze the research model that we adopted for estimation analysis is shown in Fig. 4. In Fig. 4 we report each variable relationship only in its standardized regression coefficient's sign (note that signs are consistent between the two data sets N_1 and N_2). In this model, N_T (a) represents a dependent variable as it is measured with multiple independent variables, which are N_R (a), N_B (a), and N_C (a).

5.5.2 Statistical Analysis

All statistical analyses have been performed with SPSS 20 [26]. Correlation and Regression analyses have been performed on our data set. Tables 3 reports the descriptive statistics of each variable from our dataset that we used for statistical analysis and to validate our proposed research hypotheses, as discussed in Sect. 3.

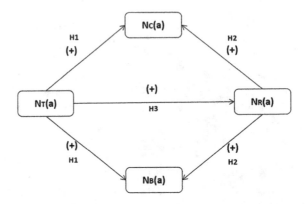

Fig. 4. Research model.

Table 3. Descriptive statistics of each variable from dataset.

	N_R (a)	N_B (a)	N_C (a)	N_T (a)
Mean	1.37	1.04	1.53	2.78
S.E of mean	0.009	0.000	0.001	0.002
S.D	10.817	0.283	1.109	1.837
Variance	117.007	0.080	1.230	3.375

Table 4 presents the correlation matrix of our data variables. Table 4 shows that correlation is significant at 0.01 level (2-tailed). All persistence variables are positively correlated with each other and, thus, have a significant impact upon each other.

Table 4. Correlation matrix of persistence variables (Pearson Index).

	N_T (a)	N_R (a)	N_B (a)	N_C (a)
N_T (a)	1	0.326	0.590	0.898
N_R (a)	0.326	1	0.254	0.235
N_B (a)	0.590	0.254	1	0.392
N_C (a)	0.898	0.235	0.392	1

The regression estimation results of the research model are shown in Table 5. All relationships between persistence metrics (i.e. N_R (a), N_B (a), and N_C (a)) and the persistence latent variable (i.e. N_T (a)) are significant, with $p < 0.001$. This confirms that factorization was performed correctly over fitted research model.

Table 5. Estimates of regression weights for the research model.

$V_{Dependent}$	$V_{Independent}$	R_W	S.E	p-value
N_R (a)	N_T (a)	0.082	0.000	< 0.001
N_B (a)	N_T (a)	0.000	0.000	< 0.001
N_C (a)	N_T (a)	0.000	0.000	< 0.001

- HYPOTHESIS H1

Hypothesis H1 "Tweeting with a high frequency of tweets is positively associated with number of topics (brands or categories) (i.e. visually *potential influencers* are the peripheral authors)" has been tested through correlation. By Table 4, both N_C (a) and N_B (a) have positive correlation of 0.898 and 0.590, respectively with N_T (a), at 0.01 level of significance. Hence, both correlation values support the hypothesis H1. It means that, *generalist* authors, who tweet about multiple topics (brands or categories), are more likely to be *content specifiers*. Such authors by having greater probability of sharing contents, can be *potential influencers* in their network.

Similarly, through visualization results we can also observe the big sized author nodes who tweet a lot about multiple brand (Fig. 3) or about multiple categories (Appendix 1).

- HYPOTHESIS H2

Similarly, hypothesis H2, "Tweeting about multiple topics (brands or categories) is positively associated with the frequency of retweets (i.e. visually, peripheral authors, connected to multiple topics, are *actual influencers*)", has been tested through correlation. By Table 4, both N_C (a) and N_B (a) have a positive correlation of 0.254 and 0.235, respectively with N_R (a), at 0.01 level of significance. Hence, both correlation values support the hypothesis H2. This means that, authors, who have a large number of retweets, are also *content specifiers* or can also be *'information sources'* or *'generators'*. Such authors can be *actual influencers* in spreading the influence among networks, as they receive large number of retweets by tweeting about multiple topics.

From a visualization standpoint, if we explore the produced graph (e.g. Fig. 3), authors who seems to be big sized nodes (visually drawn in-between multiple cluster peripheries) talking about multiple topics (brands or categories), also have a high number of retweets as well.

- HYPOTHESIS H3

Similarly, hypothesis H3, "Tweeting more frequently about a single topic (brand or category) is positively associated with the frequency of retweets (i.e. visually, authors drawn closer to single topic, are both *actual and potential influencers*)", has been tested through correlation. By observing values from Table 4, N_T (a) and N_R (a) have a positive correlation of 0.326 at 0.01 level of significance. Although the correlation coefficient is not high, the p-value in Table 5 shows significance and seems to support a positive (though weak) correlation between N_T (a) and N_R (a). As per descriptive statistics of networks, presented in Table 2, we can observe that as the number of tweets increases, the number of retweets also increases for each size or network topology.

From a visual standpoint, as shown in Fig. 3, we know that the nodes (which are drawn closer to a single brand in the innermost periphery of distinct clusters) are those authors who tweet most frequent about a specific brand in its cluster. Such authors are connected closer to cluster hubs (brands or categories), by having a high *l*-shell value as of having a high number of tweets (as discussed earlier in Sect. 4).

6 Discussion

The network layout shows that clusters are placed at a different distance from the visualization center based on the number of hubs to which they are connected. In other words, the most peripheral clusters are those in which nodes are connected to only one hub, while the central cluster is the one in which nodes are connected to the highest number of hub nodes. Within a single cluster, multiple layers seem to be formed. By implementing the *l-shell* decomposition methodology, the outside layer consists of author nodes who posted a tweet only once, as we move inward towards the brand node (hub), the frequency of tweeting increases. Hence, the closest nodes to a hub represent the authors who tweeted most about that brand and are both *potential and actual influencers*. The power-law layout algorithm has provided a network layout that is very effective in highlighting a specific property of authors which was not a measured variable in our dataset, i.e. their specificity (or generality) with respect to a topic (i.e. a brand as in Fig. 3 or category as in Fig. 5b). Authors belonging to different clusters are in fact those who are more generalist in their content sharing, since they tweet about multiple different brands. On the contrary, authors belonging to the innermost clusters are those who are very specific in sharing content related to one brand.

Since the *specificity* (generality) and *frequency of tweets* and *retweets* of authors was not an explicitly measured variable in our dataset, it is possible to posit that the proposed network layout algorithm can be considered as a powerful visual data analysis tool, since it is effective in providing visual representations of networks that help unveiling specific (implicit) properties of the represented networks. Moreover, Fig. 5 (Appendix 1) provide a few more visualizations of networks N_1 and N_2.

We also noticed that, as the graph size increases, more peripheral layers seems to be formed surrounding hub nodes, which increase the influence spread across newly formed peripheral layers in multi-layered form. Authors seem to evolve by tweeting about multiple topics among multiple peripheries. We can visually identify the increase in influence spread, as shown in Fig. 5(a), which is a larger graph of the N_1 type network, as compared to Fig. 3, where the addition of more multi-layered peripheral nodes around hub-nodes (i.e. brands) increases the influence spread across those peripheral layers. The outlier authors along the periphery can be potential influence spreaders, if they connect with other clusters through retweeting and, thus, play a critical role in determining influence. As presented in Fig. 3, network N_1 is related to the relationship between authors and brands, i.e., touristic destinations. In this case, the clustering of nodes provides a distinct clustering of those authors who have tweeted about the same destination. The layering of nodes around brands is instead related to the intensity of tweeting about a given destination; i.e., authors closer to a brand node tweet a higher number of times about that destination with respect to farther authors. The emerging semantics of the network visualization in this case is related to the *brand fidelity* of authors. The visualized network layout supports the visual analysis of those authors who have a higher fidelity to a given brand, or those authors who never tweet about that brand. Moreover, it is possible to point out which authors are tweeting about a brand as well as competing brands to support the definition of specific marketing campaigns.

7 Conclusion and Future Work

This paper proposes a novel visual approach to the analysis and exploration of social networks in order to identify and visually highlight influencers (i.e., hub nodes), and influence (i.e., spread of multi-layer peripheral nodes), represented by the opinions expressed by social media users on a given set of topics. Results show that our approach produces aesthetically pleasant graph layouts, by highlighting multi-layered clusters of nodes surrounding hub nodes (the main topics). These multi-layered peripheral node clusters represent a visual aid to understand influence.

Our approach exploits the underlying concept of power-law degree distribution with the metaphor of k-shell decomposition, thus we are able to visualize social networks in multi-layered, clustered peripheries around hub-nodes, which not only preserves the graph drawing aesthetic criteria, but also effectively represents multi-layered peripheral clusters around hub nodes. We analyzed multi-clusters, influence spread of multilayered peripheries, brand fidelity, content specificity, and potential influencers through our proposed visual framework.

Empirical testing and evaluation results show that the proposed three hypothesis that tie *content specificity, frequency of tweets* and *retweets* are supported. Moreover, the parameters like *specificity, frequency*, and *retweets* are also mutually correlated, and have a significant impact on an author's influence and encourage us to further explore social network's intrinsic characteristics. Although our experiment can be repeated with data from entities different from tourism, additional empirical work is needed to extend testing to multiple datasets and domains.

Future work will consider measures of influence with additional parameters besides frequency of sharing, content specificity and frequency of retweets. In our current work, we are studying a measure of influence, which can be used to rank influential nodes in social networks [25] and incorporate it in our research towards practical implementation.

Appendix

Figure 5 provides more visualizations of networks N_1 and N_2 of our dataset. An enlarged and zoomable version of the network Layout can be accessed online at URL http://goo.gl/WYFCjN.

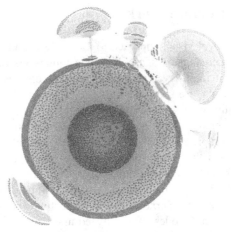

a) N_1 (Author → Brand)

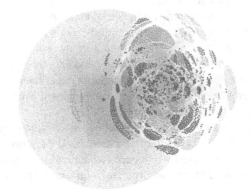

b) N_2 (Author → Category)

Fig. 5. Network visualizations of N_1 and N_2 networks.

References

1. Andersen, R., Chung, F., Lu, L.: Drawing power law graphs using a local global decomposition. Algorithmica **47**(4), 397 (2007)
2. Anholt, S.: Competitive identity: the new brand management for nations, cities and regions. Palgrave Macmillan, UK (2006)
3. Arbuckle, J.L.: IBM SPSS Amos 20 user's guide. Amos Development Corporation, SPSS Inc. (2011)
4. Asur, S., Huberman, B.A., Szabo, G., Wang, C.: Trends in social media: persistence and decay. Paper presented at the ICWSM (2011)
5. Bakshy, E., Hofman, J.M., Mason, W.A., Watts, D.J.: Everyone's an influencer: quantifying influence on twitter. Paper presented at the Proceedings of the Fourth ACM International Conference on Web Search and Data Mining (2011)

6. Barbagallo, D., Bruni, L., Francalanci, C., Giacomazzi, P.: An Empirical Study on the Relationship between Twitter Sentiment and Influence in the Tourism Domain Information and Communication Technologies in Tourism 2012, pp. 506–516. Springer, Vienna (2012)

7. Benevenuto, F., Cha, M., Gummadi, K.P., Haddadi, H.: Measuring user influence in twitter: the million follower fallacy. Paper presented at the International AAAI Conference on Weblogs and Social (ICWSM10) 2010

8. Bigonha, C., Cardoso, T.N.C., Moro, M.M., Gonçalves, M.A., Almeida, V.A.F.: Sentiment-based influence detection on twitter. J. Braz. Comput. Soc. 18(3), 169–183 (2012)

9. Boyd, D., Golde, S., Lotan, G.: Tweet, tweet, retweet: conversational aspects of retweeting on twitter. pp. 1–10. IEEE (2010)

10. Bruni, L., Francalanci, C., Giacomazzi, P., Merlo, F., Poli, A.: The relationship among volumes, specificity, and influence of social media information. Paper presented at the Proceedings of International Conference on Information Systems (2013)

11. Carmi, S., Havlin, S., Kirkpatrick, S., Shavitt, Y., Shir, E.: A model of internet topology using k-shell decomposition. Proc. Natl. Acad. Sci. 104(27), 11150–11154 (2007)

12. Chan, D.S.M., Chua, K.S., Leckie, C., Parhar, A.: Visualisation of power-law network topologies. In: The 11th IEEE International Conference on Paper presented at the Networks, ICON 2003 (2004)

13. Chang, J.-Y.: An evaluation of twitter ranking using the retweet information. J. Soc. e-Bus. Stud. 17(2), 73–85 (2014)

14. Fan, W., Gordon, M.D.: The power of social media analytics. Commun. ACM 57(6), 74–81 (2014)

15. Francalanci, C., Hussain, A.: A visual approach to the empirical analysis of social influence. Paper presented at the DATA 2014- Proceedings of 3rd International Conference on Data Management Technologies and Applications (2014). http://dx.doi.org/10.5220/0004992803190330

16. Freeman, L.C.: Centrality in social networks conceptual clarification. Soc. Netw. 1(3), 215–239 (1979)

17. Hussain, A., Latif, K., Rextin, A., Hayat, A., Alam, M.: Scalable visualization of semantic nets using power-law graphs. Appl. Math. Inf. Sci. 8(1), 355–367 (2014)

18. Hutto, C.J., Yardi, S., Gilbert, E.: A longitudinal study of follow predictors on twitter. Paper presented at the Proceedings of the SIGCHI Conference on Human Factors in Computing Systems (2013)

19. Kitsak, M., Gallos, L.K., Havlin, S., Liljeros, F., Muchnik, L., Stanley, H.E., Makse, H.A.: Identification of influential spreaders in complex networks. Nat. Phys. 6(11), 888–893 (2010)

20. Klotz, C., Ross, A., Clark, E., Martell, C.: Tweet!– And I Can Tell How Many Followers You Have Recent Advances in Information and Communication Technology, pp. 245–253. Springer, Heidelberg (2014)

21. Kwak, H., Lee, C., Park, H., Moon, S.: What is twitter, a social network or a news media? Paper presented at the Proceedings of the 19th International Conference on World Wide Web (2010)

22. Leavitt, A., Burchard, E., Fisher, D., Gilbert, S.: The influentials: new approaches for analyzing influence on twitter. Web Ecol. Proj. 4(2), 1–18 (2009)

23. Meraz, S.: Is there an elite hold? Traditional media to social media agenda setting influence in blog networks. J. Comput.-Mediated Commun. 14(3), 682–707 (2009)

24. Metra, I.: Influence based exploration of twitter social network. Politecnico di Milano, Milan, Italy (2014)

25. Myers, S.A., Leskovec, J.: The bursty dynamics of the twitter information network. Paper presented at the Proceedings of the 23rd International Conference on World Wide Web (2014)
26. Pallant, J.: SPSS Survival Manual: A Step by Step Guide to Data Analysis using SPSS. McGraw-Hill International, New York (2010)
27. Perline, R.: Strong, weak and false inverse power laws. Stat. Sci. **20**(1), 68–88 (2005)
28. Ren, Z.-M., Zeng, A., Chen, D.-B., Liao, H., Liu, J.-G.: Iterative resource allocation for ranking spreaders in complex networks. EPL (Europhys. Lett.) **106**(4), 48005 (2014)
29. Sparrowe, R.T., Liden, R.C., Wayne, S.J., Kraimer, M.L.: Social networks and the performance of individuals and groups. Acad. Manag. J. **44**(2), 316–325 (2001)

Using Behavioral Data Mining to Produce Friend Recommendations in a Social Bookmarking System

Matteo Manca, Ludovico Boratto$^{(\boxtimes)}$, and Salvatore Carta

Dipartimento di Matematica e Informatica, Università di Cagliari,
Via Ospedale 72, 09124 Cagliari, Italy
{matteo.manca,ludovico.boratto,salvatore}@unica.it

Abstract. Social recommender systems have been developed to filter the large amounts of data generated by social media systems. A type of social media, known as *social bookmarking system*, allows the users to tag bookmarks of interest and to share them. Although the popularity of these systems is increasing and even if users are allowed to connect both by following other users or by adding them as friends, no friend recommender system has been proposed in the literature. Behavioral data mining is a useful tool to extract information by analyzing the behavior of the users in a system. In this paper we first perform a preliminary analysis that shows that behavioral data mining is effective to discover how similar the preferences of two users are. Then, we exploit the analysis of the user behavior to produce friend recommendations, by analyzing the resources tagged by a user and the frequency of each used tag. Experimental results highlight that, by analyzing both the tagging and bookmarking behaviors of a user, our approach is able to mine preferences in a more accurate way with respect to a state-of-the-art approach that considers only the tags.

Keywords: Social bookmarking · Friend recommendation · Behavioral data mining · Tagging system

1 Introduction

The explosion of the Web 2.0 led to a continuous growth of information sources, with daily uploaded resources shared by many users. Social media systems are web-based services that allow individuals to construct a public or semi-public profile within a bounded system, create a list of other users with whom they share a connection, and view and traverse their list of connections and those made by other users within the system [13]. Moreover, Guy et al. [21] highlight that a

This work is partially funded by Regione Sardegna under project SocialGlue, through PIA - Pacchetti Integrati di Agevolazione "Industria Artigianato e Servizi" (annualità 2010).

M. Helfert et al. (Eds.): DATA 2014, CCIS 178, pp. 99–116, 2015.
DOI: 10.1007/978-3-319-25936-9_7

social media system is characterized by a user-centered design, user-generated content (e.g., tags), social networks, and online communities.

Social Interaction Overload Problem. The growth of the user population and the large amount of content in these system lead to the well-known "social interaction overload" problem [22,35]. Social interaction overload is related to the excessive amount of users and items that each user can interact with. This leads to the scarcity of attention, which does not allow the users to focus on users or items that might be interesting for her/him. In order to filter information in the social media systems domain, in the last few years the research on recommendation has brought to the development of a new class of systems, named *social recommender systems* [34]. These systems face the social interaction overload problem, by suggesting users or items that users might be interested in. In particular, user recommendation in a social domain aims at suggesting *friends* (i.e., recommendations are built for pairs of users that are likely to be interested in each other's content) or *people to follow* (i.e., recommendations are built for a user, in order to suggest users that might be interesting for her/him) [22].

Social Bookmarking Systems and Recommendation. A *social bookmarking system* is a form of social media, which allows users to use keywords (*tags*) to describe resources that are of interest for them, helping to organize and share these resources with other users in the network [19]. The most widely-known examples of social bookmarking systems are Delicious[1], where the bookmarked resources are web pages, CiteULike[2], where users bookmark academic papers, and Flickr[3], where each picture can be annotated with tags.

Even if users are connected in a social network and interact with each other, and the use of these systems is widespread (in 2014, one million photos per day have been shared on Flickr[4]), to the best of the authors' knowledge, no approach in the literature recommends friends in a social bookmarking system.

Behavioral data mining is the process of extracting information from data by analyzing the behavior of the users in the system. Its effectiveness has been validated in various areas, such as the detection of tag clusters [10,12], the creation of web personalization services [30], and the improvement of web search ranking [1].

Our Contributions. In this paper, we present a friend recommender system in the social bookmarking domain. By mining the content of the target user, the system recommends users that have similar interests. More specifically, the problem statement is the following:

Problem 1. We are given a social bookmarking system, defined as a tuple $Q = \{U, R, T, A, C\}$, where:

- U, R, and T are sets of *users*, *resources*, and *tags*;
- A is a ternary relation between the sets of *users*, *resources*, and *tags*, i.e., $A \subseteq U \times R \times T$, whose elements are the tag *assignments* of a user for a resource;
- C is a binary relation between the users, i.e., $C \subseteq U \times U$, whose elements express the *connection* among two users. If we represent the user social relations by means of a graph, in which each node represents a user $u \in U$ and each edge $c \in C$ represents a connection among two users, we will have an undirected edge if the users are connected as *friends* and a directed edge if one user *follows* the other.

Our aim is to define a function $f : U \times U \rightarrow C$, which defines if, given two users $u \in U$ and $m \in U$, there is a undirected connection $c \in C$ among them.

In order to motivate our proposal and design our system, we first perform a novel study of the user behavior in a social bookmarking system.

The scientific contributions coming from our proposal are the following:

- we propose an analysis of the user behavior in a social bookmarking system, in order to present the motivation that leads to the design of a friend recommender system based on behavioral data mining;
- we propose the first system in the literature that recommends friends in this domain (other approaches in the literature recommend people to follow but, as previously highlighted, this is a different research topic);
- we study how to mine content in this context, i.e., what information should be used to produce the recommendations and which importance should the different types of content have in the recommender system.

This paper extends the work presented in [28]. With respect to our original work, the presentation has been completely reorganized, in order to integrate the following extensions: (i) the contextualization with the state of the arthas been enhanced, in order to include recent related work; (ii) we present a novel study of the user behavior in a social bookmarking system, in order to capture similarities between the users that can be exploited to produce friend recommendations; (iii) we provide additional details on the algorithms that compose our system.

The proposed approach, thanks to its capability to exploit the interests of the users and being the first developed in this domain, puts the basis to a research area not previously explored by the existing social recommender systems.

The rest of the paper is organized as follows: Sect. 2 presents related work; Sect. 3 describes the details of the recommender system presented in this paper: starting from an analysis of the user behavior in a social bookmarking system, we design our friend recommender system, and present the algorithms that compose it; Sect. 4 illustrates the conducted experiments and outlines main results; Sect. 5 contains comments, conclusions, and future work.

2   Related Work

This section presents related work on user recommendation in the social domain. These systems can be classified into three categories, based on the source of data used to build the recommendations:

1. Systems based on the analysis of social graphs, which explore the set of people connected to the target user, in order to produce recommendations. These systems recommend either the closest users in the graph, like friends of friends and followees of followees (the "People you may know" feature offered by Facebook [33] is the most widely known example of this approach), or recommend the users that have the highest probability to be crossed in a random walk of the social graph (the main reference for this type of systems is the "Who to follow" recommendation in Twitter [20]).
2. Systems that analyze the interactions of the users with the content of the system (tags, likes, posts, etc.). In order to exploit the user interests, these systems usually build a user profile by giving a structured form to content, thanks to the use of metrics like TF-IDF. An example of this class of systems is presented in [18].
3. Hybrid systems, which consider both the social graph and the interactions of the users with the content (an example is represented by [24]).

The rest of the section presents the main approaches for each class of systems.

2.1   Systems Based on the Analysis of Social Graphs

Barbieri et al. [3] recently presented an approach to predict links between users with a stochastic topic model. The model also represents whether a connection is "topical" or "social" and produces an explanation of the type of recommendation produced.

In [20], the authors present Twitter's user recommendation service, which allows the users to daily create a huge amount of connections between users that share common interests, connections, and other factors. In order to perform the recommendations, the authors build a Twitter graph in which the vertices represent the users and the directed edges represent the "follow" relationship. The graph is stored in a graph database called FlockDB, and then data are processed with Cassovary (an open source in-memory graph processing engine). The system builds the recommendations by means of a user recommendation algorithm for directed graphs based on SALSA. In the next section, we are going to analyze this system, in order to design our proposal.

In [26] the authors model the user recommendation problem as a link prediction problem. They develop several approaches, which analyze the proximity of nodes in the graph of a social network in order to infer the probability of new connections among users. Experiments show that the network topology is a good tool to predict future interactions.

Arru et al. [2] propose a user recommender system for Twitter, based on signal processing techniques. The presented approach defines a pattern-based similarity function among users and uses a time dimension in the representation of the user profiles. Our system is different, because we aim at suggesting friends while on Twitter there is no notion of "friend" but it works with "people to follow".

2.2 Systems Based on the Interactions with the Content

Quercia et al. [32] describe a user recommender system based on collocation. The proposed framework, called FriendSensing, recommends friends by analyzing collocation data. In order to produce the recommendations, the system uses geographical proximity and link prediction theories. In our domain we do not have such a type of information, so we cannot compare with this algorithm.

In [15], the researchers present a study that considers different features in a user profile, behavior, and network, in order to explore the effect of *homophily* on user recommendations. They use the Dice coefficient on two users sets of tags and they find that similar tags do not represent a useful source of information for link prediction, while mutual followers are more useful for this purpose. As previously highlighted, the presented friend recommender system focuses on producing friend recommendations based on users' content (tag, bookmarks, etc.).

2.3 Hybrid Systems

In [37], the authors propose a framework for user recommendation, based on user' interests and tested on Yahoo! Delicious. The proposed framework operates in two main steps: first, it models the user' interests by means of a tag based graph community detection and represents them with a discrete topic distribution; then, it uses the Kullback-Leibler divergence function to compute the similarity between users' topic distribution and the similarity values are used to produce interest-based user recommendations. Differently from this framework, the aim of the approach proposed in this paper is to produce friend recommendations (i.e., bidirectional connections) and not unidirectional user recommendations.

Chen et al. [18] present a people recommender system in an enterprise social network called Beehive, designed to help users to find known, offline contacts and discover new friends on social networking sites. With the proposed study, the authors demonstrate that algorithms that use similarity of user-created content were stronger in discovering new friends, while algorithms based on social network information were able to produce better recommendations.

In [24], the authors propose a user recommender system (called *Twittomender*) that, for each user, builds a user profile based on user's recent Twitter activity and user's social graph. The proposed system operates in two different manners; in the former mode the user puts a query and the system retrieves a ranked list of users, while in the latter mode the query is automatically generated by the system and it is mined by the user profile of the target user (the target user is the one that receives the recommendations). Our proposal does not use the social graph and, furthermore, in building recommendations it considers the friendship relationship and not the "user to follow" relationship.

In [23], the authors present a recommender system for the IBM Fringe social network, based on aggregated enterprise information (like org chart relationships, paper and patent co-authorships, etc.) retrieved using SONAR, which is a system that collects and aggregates these types of information. The authors deployed the people recommender system as a feature of the social network site and the results showed a highly significant impact on the number of connections on the site, as well as on the number of users who invite others to connect.

3   Friend Recommendation by Mining User Behavior

3.1   User Behavior in a Social Bookmarking System

This section aims at analyzing the user behavior in a social bookmarking system from a friend recommendation point of view. In particular, we study how the bookmarking activity of a user is related to that of the others.

This analysis has been conducted on a Delicious dataset, distributed for the HetRec 2011 workshop [17], which contains:

- 1867 users;
- 69226 URLs;
- 53388 tags;
- 7668 bi-directional user relations;
- 437593 tag assignments (i.e., tuples [user, tag, URL]);
- 104799 bookmarks (i.e., distinct pairs [user, URL]).

In their profiles, users had an average of 123.697 tags used to bookmark the resources, and an average of 56.132 bookmarked resources.

Definition 1. Let P_T be a random variable that expresses the number of tags in common between two users, and let P_R be a random variable that expresses the number of resources in common between two users.

The first part of the study aims at measuring the cumulative distribution function of P_T, which will allow us to evaluate the probability for two users to have at least t tags in common:

$$F_{P_T}(t) = P(P_T \geq t) \tag{1}$$

Figure 1 shows the value of the function for all the possible amounts of common tags between a pair of users. As it can be noticed, the values of the function F_{P_T} follow a power law distribution. Results show that in a social bookmarking system users have a similar behavior in the use of tags. Indeed, we can notice that almost 50 % of the pairs of users in the system share at least 4 tags. Moreover, we can notice that almost 3 % of the users have 40 tags in common, which indicate a very similar behavior in their tagging activity.

In the second part of our behavioral analysis we measure the cumulative distribution function of P_R, which will allow us to evaluate the probability for two users to have at least r resources in common:

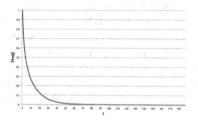

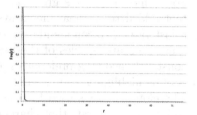

Fig. 1. Probability for two users to have at least t tags in common.

Fig. 2. Probability for two users to have at least r resources in common.

$$F_{P_R}(r) = P(P_R \geq r) \tag{2}$$

Figure 2 shows the value of the function for all the possible amounts of common resources between a pair of users. Again, the values of the function F_{P_R} follow a power law distribution. These results highlight a very different trend, which shows that more than 2 % of the users have at least one tag in common, with a sudden drop in the results. Even if the probability values are very low, we can still notice that the absolute numbers report us interesting results, such as more than 1000 pairs of users with at least 16 bookmarked resources in common.

This analysis allows us to infer some properties related to the user behavior in a social bookmarking system, recapped below:

- the behavior of two users in a social bookmarking system is related, both in the way they use tags and in the resources they bookmark with those tags;
- the use of tags represents a stronger form of connection, with respect to the amount of resources in common between two users. This happens because a user classifies a resource with more tags, so the probability that two users use the same tags is higher than the one to bookmark the same resource.

This behavioral analysis has been one of the aspects that characterized the design of the system, which is presented next.

3.2   System Design

The objective of our work is to build a friend recommender system in the social bookmarking domain. In its design, we considered the following aspects:

(a) In [20], the authors highlight that Twitter is an "interest graph", rather than a "social graph". A problem highlighted in the paper is that the analysis of such a graph suffers from scalability issues and, in order to contain the complexity of the recommender system, no personal information can be used to produce the recommendations. The definition of interest graph can also be extended to social bookmarking systems, since a user can add as a friend or follow another user, in order to receive her/his newly added bookmarks.

(b) Social media systems grow rapidly. This means that the amount of content added to a social media system and the user population increase at a fast rate. A recommender system that operates in this context needs to build accurate profiles of the users, which have to be up-to-date with the constantly evolving preferences of the users.

(c) As [37] highlights, the tagging activity of the users reflects their interests. Therefore, the tags used by a user are an important source of information to infer the interests that characterize them.

Taking into account all these aspects, we designed a recommender system that operates in the following way.

Regarding point (a), we designed a system that only analyzes the content of the users (i.e., the tagged bookmarks). So, in order to avoid the limitations related to the graph analysis in this domain, our system belongs to the second class presented in Sect. 2, i.e., the one that analyzes the interactions of the users with the content of the system.

Regarding point (b), in order to efficiently and quickly update user profiles, our system computes user similarities with low computational cost metrics, which exploit the set of resources used by each user and the tags used to classify them.

Regarding point (c), we embraced the theory that user interest is reflected by the tagging activity and we extended it, by following the intuition that users with similar interests make a similar use of tags and bookmark the same resources.

Reader can refer to [27] for a detailed analysis of a friend recommender system in this domain from the design and architectural points of view.

A description of the system is presented next.

3.3   Algorithms

Given a target user $u_t \in U$, the system recommends the users with a high tag-based user similarity and a high percentage of common resources. The system works in five steps:

1. *Tag-based user profiling.* Given the tag assignments of each user, this step builds a user profile, based on the frequencies of the tags used by a user.
2. *Resource-based user profiling.* Given the tag assignments of each user, this step builds a user profile, based on the resources bookmarked by a user.
3. *Tag-based similarity computation.* The first metric, calculated among a target user u_t and the other users, is based on the tag-based user profile. Pearson's correlation is used to derive the similarity.
4. *User interest computation.* The second computed metric is the interest of a user towards another user and it is represented by the percentage of common resources among them.
5. *Recommendations selection.* This step recommends to u_t the users with both a tag-based and a user interest higher than a threshold value.

In the following, we will give a detailed description of each step.

Tag-Based User Profiling. This step builds a user profile, based on the tags available in the tag assignments of a user, considering the frequency of each used tag. Given the sets defined in Problem 1, we can first consider the tag assignments of a user u as follows:

Definition 2. Let $A(u) \subseteq A$, be the subset of A, whose elements are the triples that contain a user $u \in U$, i.e., $\forall r \in R \wedge \forall t \in T, (u, r, t) \in A \Rightarrow (u, r, t) \in A(u)$. Let $A(u, t) \subseteq A(u)$, be the subset of A(u), whose elements are all the triples that contain a tag $t \in T$ used by a user $u \in U$, i.e., $\forall r \in R, (u, r, t) \in A(u) \Rightarrow (u, r, t) \in A(u, t)$.

A user can be profiled, according to her/his use of the tags, by considering the relative frequency of each tag, as follows:

$$v_{uj} = \frac{|A(u, t_j)|}{|A(u)|} \tag{3}$$

Equation 3 estimates the importance of a tag $t_j \in T$ in the profile of a user $u \in U$, by defining the relative frequency as the number of times the tag t_j was used, divided by the number of tag assignments of u.

A tag-based user profile can be implemented by representing each user $u \in U$ as a vector $\vec{v_u} = \{v_{u1}, v_{u2}, ..., v_{uk}\}$, where each element v_{uj} is the previously defined relative frequency and k is the number of tags in the system.

Resource-Based User Profiling. This step builds another user profile, based on the resources bookmarked by each user. A user can be profiled, according to her/his bookmarked resources, by considering the fact that she/he bookmarked a resource (i.e., she/he expressed interest in it):

$$v_{uj} = \begin{cases} 1 \text{ if } \exists t \in T \mid (u, r_j, t) \in A(u) \\ 0 \text{ otherwise} \end{cases} \tag{4}$$

Equation 4 estimates the interest of a user u in a resource r_j with a binary value, equal to 1 in case r_j was bookmarked by u, and 0 otherwise.

A resource-based user profile can be implemented by representing each user $u \in U$ by means of a binary vector $\vec{v_u} = \{v_{u1}, v_{u2}, ..., v_{un}\}$, which represents the resources tagged by each user. Each element v_{uj} is defined as previously illustrated and n is the number of resources in the system.

Tag-Based Similarity Computation. Since in [37] the authors highlight that the interests of the users are reflected in their tagging activities, our system computes the similarity among two tag-based user profiles with the Pearson's correlation coefficient [31]. This metric was chosen because, as proved by Breese et al. [14], it is the most effective for the similarity assessment among users. Moreover, an efficient algorithm that exploits a support-based upper bound exists [36].

Let (u, m) be a pair of users represented respectively by vectors $\vec{v_u}$ and $\vec{v_m}$. Our algorithm computes the tag-based user similarity ts as defined in Eq. 5:

$$ts(u,m) = \frac{\sum_{j \subset T_{um}} (v_{uj} - \overline{v}_u)(v_{mj} - \overline{v}_m)}{\sqrt{\sum_{j \subset T_{um}} (v_{uj} - \overline{v}_u)^2} \sqrt{\sum_{j \subset T_{um}} (v_{mj} - \overline{v}_m)^2}} \quad (5)$$

where T_{um} represents the set of tags used by both users u and m and values $\overline{v}_u$ and $\overline{v}_m$ represent, respectively, the mean of the frequencies of user u and user m. This metric compares the frequencies of all the tags used by the considered users. The similarity values range from 1.0, which indicates complete similarity, to -1.0, which indicates complete dissimilarity. Herlocker et al. [25] demonstrated that negative similarities are not significant to evaluate the correlation among users, so in our algorithm we consider only positive values.

User Interest Computation. Given a pair of users (u, m), in this step, we compute two metrics based on the resources tagged by users. The former, $ui(u, m)$, represents the interest of the user u towards user m, while the latter, $ui(m, u)$, represents the interest of the user m toward the user u.

We first consider the set of resources bookmarked by each user, and then consider the resources in common between two users.

Definition 3. Let $R(u) \subseteq R$ be the subset of resources used by a user $u \in U$, i.e., $\forall r \in R, (u, r, t) \in A(u) \Rightarrow r \in R(u)$. Let $D(u, m) = R(u) \cap R(m)$ be the subset of resources bookmarked by both user u and user m.

The *user interest* of a user u in a user m can be estimated as:

$$ui(u, m) = \frac{|D(u, m)|}{|R(u)|} \quad (6)$$

The level of interest of a user u in a user m is estimated as the number of resources bookmarked by both the users, divided by the number of resources bookmarked by user u. This means that the interest of the user m in user u depends on the number of resources bookmarked by m (i.e., when calculating $ui(m, u)$, the denominator would be $|R(m)|$).

The previously defined user interest ui, can be implemented, by using the two resource-based user profiles $\overrightarrow{v_u}$ and $\overrightarrow{v_m}$, as follows:

$$ui(u, m) = \frac{\sum_{j=1}^{n} v_{uj} v_{mj}}{\sum_{j=1}^{n} v_{uj}} * 100 \quad (7)$$

$$ui(m, u) = \frac{\sum_{j=1}^{n} v_{uj} v_{mj}}{\sum_{j=1}^{n} v_{mj}} * 100 \quad (8)$$

where n is the total number of resources of the system.

Recommendations Selection. As Marlow et al. highlight [29], the use of tags and resources in a social tagging system is associated to two different types of behavior in a tagging system. Therefore, the aggregation of the tag-based similarity and of the user interests into a single score would blur the information on how similar two users are for each type of behavior. This could lead to potentially inaccurate friend recommendations, like two users that use the same tags to describe completely different and unrelated types of resources. Therefore, our recommendation algorithm filters the users by considering both types of behavior. Once the tag-based similarities and the user interests have been computed for each pair of users, our system choses a set of users to recommend to the target user by selecting:

- the ones that have a tag-based user similarity higher than a threshold value α (i.e., $ts > \alpha$);
- the ones that have a user interest (at least one of the two computed) higher than a threshold value β (i.e., $ui > \beta$).

Definition 4. Given a target user u_t, the candidate set of users to recommend $S(u_t)$ can be defined as

$$S(u_t) = \{u_i \in U \,|\, ts(u_t, u_i) > \alpha \,\&\&\, (ui(u_t, u_i) > \beta) \,||\, (ui(u_i, u_t) > \beta)\} \qquad (9)$$

As Eq. 9 shows, the system creates a recommendation between two users if a similarity on both types of behavior exists. In particular, since in Sect. 3 we showed that the user interest (i.e., the amount of common resources between two users) represents a weaker form of connection, we relaxed the constraint on the reciprocity of the user interest and compared the similarities with an OR operator.

4 Experimental Framework

This section presents the framework used to perform the experiments.

4.1 Dataset and Pre-processing

Experiments were conducted on a Delicious dataset distributed for the HetRec 2011 workshop [17]. It contains:

- 1867 users, which represent the elements of the set U previously defined;
- 69226 URLs, which represent the elements of the set R previously defined;
- 53388 tags, which represent the elements of the set T previously defined;
- 7668 bi-directional user relations, which represent the elements of the relation C previously defined;
- 437593 tag assignments (i.e., the tuples $(user, tag, URL)$), which represent the elements of the relation A previously defined;
- 104799 bookmarks (i.e., the distinct pairs $(user, URL)$), which represent the elements of the union of the subsets $R(u)$ previously defined.

We pre-processed the dataset, in order to remove the users that were considered as "inactive", i.e., the ones that used less than 5 tags or less then 5 URLs.

4.2 Metrics

This section presents the metrics used for the performance evaluation of the system.

Precision. In order to measure the accuracy of the system, we evaluate the effectiveness of the recommendations (i.e., which recommended friends are actually friends with the target user), by measuring its *precision*.

Definition 5. Let W be the total amount of recommendations produced by the system, i.e., $W = \cup S(u_t), \forall u_t \in U$. This set represents the positive outcomes, i.e., the sum of the *true positive* and the *false positive* recommendations. Let Z be the amount of correct recommendations produced by the system, i.e., $Z \subseteq W = \{(u, m) \mid (u, m) \in W \wedge (u, m) \in C\}$. So, Z represents the subset of recommendations for which there is a relation (i.e., a friend correlation) in the dataset. This subset represents the *true positive* recommendations.

Given the previously defined two sets, we can measure the *precision* of our recommender system as the number of correct recommendations, divided by the number of produced recommendations:

$$precision = \frac{true\ prositive}{true\ prositive\ +\ false\ positive} = \frac{|Z|}{|W|} \tag{10}$$

Even if the recall metric is usually computed along with precision, it captures a perspective that differs from the way our system operates. We propose a constraint-based approach that reduces the amount of selected users, while recall measures completeness and quantity of recommendations [16]. Because of the nature of the metric, it would be misleading to compute it in order to evaluate the accuracy of our system.

Percentage of Satisfied Users. This metric evaluates the system from a similar (but different) perspective with respect to its precision. In fact, precision measures for how many couples of users a correct recommendation was produced, while the *percentage of satisfied users* measures for how many individual users a correct recommendation was produced.

Definition 6. Let $X \subseteq U$ be the subset of users for which a recommendation was produced, i.e., $X = \{u \in U | \exists (u, m) \in W\}$. Let $Y \subseteq U$ be the subset of users for which a correct recommendation was produced, i.e., $Y = \{u \in U | \exists (u, m) \in Z\}$.

The percentage of users satisfied by the recommendations can be computed by dividing the set of users for which a correct recommendations was produced by the set of users for which a recommendation was produced, as follows:

$$\%\ satisfied\ users = \frac{|Y|}{|X|} * 100. \tag{11}$$

4.3   Strategy

We performed two different experiments. The first aims to make an *evaluation of the recommendations*, by measuring the precision of the system with different threshold values. The second experiment, makes an *evaluation of the satisfied users* in the produced recommendations, given a precision value.

In order to evaluate the recommendations, we compare our approach with a state-of-the-art policy [37]. Zhou et al. [37] developed a tag-based user recommendation framework and demonstrated that tags are the most effective source of information to produce recommendations. We compare the performance of our system with respect to that of the reference one (which uses only the tags, i.e., $ui = 0$), in terms of precision. Supported by the thesis that the use of only one source of data leads to a better performance, we considered a second reference system, which exploits only the user interest (i.e., $ts = 0$).

During the analysis of the performance, we evaluated all the values of the parameters α and β between 0 and 1, using a 0.1 interval.

4.4 Experiments

The details of each performed experiment and its results are now presented.

Evaluation of the Recommendations. Given a target user u_t, the system builds a candidate set, $S(u_t)$, of users to recommend. For each recommended user $u_i \in S(u_t)$, we analyze the bi-directional user relations in the dataset (i.e., if $(u_t, u_i) \in C$), to check if there is a connection between the target user u_t and the recommended user u_i (i.e., if the users are friends). This experiment analyzes the performance of the system in terms of *precision*. Given different values of α and β, the precision of the system is calculated, in order to analyze how the performance of the system vary as the similarity between users grows. The results are illustrated in Figs. 3 and 4.

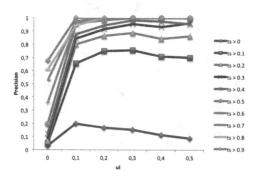

Fig. 3. Precision of the system with respect to user interest ui (Color figure online).

Figure 3 shows how the precision values change with respect to the user interest ui. The figure contains a line for each possible value α of the threshold for the tag-based user similarity ts. We can observe that the precision values grow proportionally to the ui values. This means that the more similar the users (both in terms of tag-based similarity and of user interest), the better the system performs. However, for ui values higher than 0.5 no user respects the constraints, so the system cannot produce any recommendation.

Figure 4 shows the same results from the tag-based user similarity point of view. The figure presents the precision values, with respect to the tag-based user similarity ts; here, each line shows the results for a given value β of the threshold for the user interest ui. Also from this perspective, the precision grows proportionally to ts.

The blue lines in Figs. 3 and 4 show the results of the reference systems, where $ts = 0$ and $ui = 0$. In both cases, the combined use of the two metrics improves the quality of the recommendations with respect to the cases where only one is used.

Evaluation of the Satisfied Users. The second experiment aims at analyzing the trend of the satisfied users, with respect to the precision values. So, for each precision value obtained in the previous experiment, we computed the percentage of satisfied users as shown in Eq. 11. In order to present the results, Fig. 5 reports just a subset of precision values. These values have been selected dividing the range $[0 - 1]$ of possible precision values into intervals of 0.1 (i.e., $[0 - 0.1)$, $[0.1 - 0.2)$, ..., $[0.9 - 1])$ and assigning each previously computed value of precision to the right interval. From each interval, we selected the record that corresponds to the precision value that led to the maximum percentage of satisfied users. The reason why there are no values for the intervals $[0.2 - 0.3)$ and $[0.4 - 0.5)$, is that in the previous experiments there are no values of α and β that led to precision values inside these intervals.

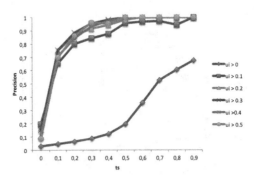

Fig. 4. Precision of the system with respect to tag-based user similarity ts (Color figure online).

In Fig. 5 we can see that the percentage of satisfied users grows as the precision grows. Given that also in the previous experiments we obtained that the more similar the users were, the higher the precision was, we can conclude that

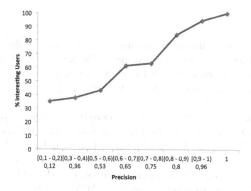

Fig. 5. Percentage of satisfied users for different values of precision.

the more similar the users are (both in terms of tag-based similarity and of user interest), the higher is the likelihood to be satisfied by the recommendations.

These results show an interesting property of our recommender system. In fact, even if the precision values are split into intervals that cover the same range (i.e., 0.1), there are two of them (i.e., $[0.6 - 0.7)$ and $[0.8 - 0.9)$) in which the percentage of individual users satisfied by the recommendations significantly increases. So, this experiment, by showing the impact of precision on individual users, is very useful in order to tune the parameters of the system.

5   Conclusions

In this paper we proposed a novel friend recommender system that operates in the social bookmarking domain. We started our study with an analysis of the effectiveness of behavioral data mining to derive similarities between users; this highlighted that users share a similar behavior, both in the way they tag and in the resources they bookmark. This led to the design and the development of a friend recommender system that exploited both the common tags and resources between the users in order to predict connections between them. The results show the capability of our approach at producing friend recommendations, both in terms of precision and of the amount of satisfied users.

Since in the last few years approaches to produce group recommendations in contexts in which groups are not available have been developed [4–9,11], future work will study our approach in this context. The analysis of the behavior will allow us to derive connections between the users using behavioral features, form the groups of users, and produce recommendations to them.

References

1. Agichtein, E., Brill, E., Dumais, S.: Improving web search ranking by incorporating user behavior information. In: Proceedings of the 29th Annual International ACM SIGIR Conference on Research and Development in Information Retrieval, SIGIR 2006, pp. 19–26. ACM, New York (2006). http://doi.acm.org/10.1145/1148170. 1148177
2. Arru, G., Gurini, D. F., Gasparetti, F., Micarelli, A., Sansonetti, G.: Signal-based user recommendation on twitter. In: Carr, L., Laender, A.H.F., Lóscio, B.F., King, I., Fontoura, M., Vrandecic, D., Aroyo, L., de Oliveira, J.P.M., Lima, F., Wilde, E. (eds.) 22nd International World Wide Web Conference, WWW 2013, 13–17 May 2013, Rio de Janeiro, Brazil, Companion volume, pp. 941–944. International World Wide Web Conferences Steering Committee/ACM (2013)
3. Barbieri, N., Bonchi, F., Manco, G.: Who to follow and why: link prediction with explanations. In: Proceedings of the 20th ACM SIGKDD International Conference on Knowledge Discovery and Data Mining, KDD 2014, pp. 1266–1275. ACM, New York (2014). http://doi.acm.org/10.1145/2623330.2623733
4. Boratto, L., Carta, S.: State-of-the-art in group recommendation and new approaches for automatic identification of groups. In: Soro, A., Vargiu, E., Armano, G., Paddeu, G. (eds.) Information Retrieval and Mining in Distributed Environments. SCI, vol. 324, pp. 1–20. Springer, Heidelberg (2011)
5. Boratto, L., Carta, S.: Impact of content novelty on the accuracy of a group recommender system. In: Bellatreche, L., Mohania, M.K. (eds.) DaWaK 2014. LNCS, vol. 8646, pp. 159–170. Springer, Heidelberg (2014). http://dx.doi.org/10.1007/978-3-319-10160-6
6. Boratto, L., Carta, S.: Modeling the preferences of a group of users detected by clustering: a group recommendation case-study. In: Proceedings of the 4th International Conference on Web Intelligence, Mining and Semantics (WIMS14), WIMS 2014, pp. 16:1–16:7. ACM, New York (2014). http://doi.acm.org/10.1145/2611040. 2611073
7. Boratto, L., Carta, S.: The rating prediction task in a group recommender system that automatically detects groups: architectures, algorithms, and performance evaluation. J. Intell. Inf. Syst., 1–25 (2014). http://dx.doi.org/10.1007/s10844-014-0346-z
8. Boratto, L., Carta, S.: Using collaborative filtering to overcome the curse of dimensionality when clustering users in a group recommender system. In: Proceedings of 16th International Conference on Enterprise Information Systems (ICEIS), pp. 564–572 (2014)
9. Boratto, L., Carta, S., Chessa, A., Agelli, M., Clemente, M. L.: Group recommendation with automatic identification of users communities. In: Proceedings of the 2009 IEEE/WIC/ACM International Joint Conference on Web Intelligence and Intelligent Agent Technology, WI-IAT 2009, vol. 03, pp. 547–550. IEEE Computer Society, Washington, DC (2009). http://dx.doi.org/10.1109/WI-IAT.2009.346
10. Boratto, L., Carta, S., Manca, M., Mulas, F., Pilloni, P., Pinna, G., Vargiu, E.: A clustering approach for tag recommendation in social environments. Int. J. E Bus. Dev. **3**, 126–136 (2013)
11. Boratto, L., Carta, S., Satta, M.: Groups identification and individual recommendations in group recommendation algorithms. In: Picault, J., Kostadinov, D., Castells, P., Jaimes, A. (eds.) Practical Use of Recommender Systems, Algorithms and Technologies 2010. CEUR Workshop Proceedings, vol. 676, November 2010. http://ceur-ws.org/Vol-676/paper4.pdf

12. Boratto, L., Carta, S., Vargiu, E.: RATC: a robust automated tag clustering technique. In: Di Noia, T., Buccafurri, F. (eds.) EC-Web 2009. LNCS, vol. 5692, pp. 324–335. Springer, Heidelberg (2009)

13. Boyd, D.M., Ellison, N.B.: Social network sites: definition, history, and scholarship. J. Comput. Mediated Commun. **13**(1), 210–230 (2007)

14. Breese, J.S., Heckerman, D., Kadie, C.: Empirical analysis of predictive algorithms for collaborative filtering. In: Proceedings of the Fourteenth Conference on Uncertainty in Artificial Intelligence, UAI 1998, pp. 43–52. Morgan Kaufmann Publishers Inc., San Francisco (1998). http://dl.acm.org/citation.cfm?id=2074094.2074100

15. Brzozowski, M.J., Romero, D.M.: Who should i follow? recommending people in directed social networks. In: Adamic, L.A., Baeza-Yates, R.A., Counts, S. (eds.) Proceedings of the Fifth International Conference on Weblogs and Social Media, 17-21 July 2011. The AAAI Press, Barcelona (2011)

16. Buckland, M., Gey, F.: The relationship between recall and precision. J. Am. Soc. Inf. Sci. **45**(1), 12–19 (1994). http://dx.doi.org/10.1002/(SICI)1097-4571(199401)45:1⟨12:AID-ASI2⟩3.0.CO;2-L

17. Cantador, I., Brusilovsky, P., Kuflik, T.: Second workshop on information heterogeneity and fusion in recommender systems (hetrec2011). In: Mobasher, B., Burke, R.D., Jannach, D., Adomavicius, G. (eds.) Proceedings of the 2011 ACM Conference on Recommender Systems, RecSys 2011, 23–27 October 2011, Chicago, IL, USA, pp. 387–388. ACM (2011)

18. Chen, J., Geyer, W., Dugan, C., Muller, M.J., Guy, I.: Make new friends, but keep the old: recommending people on social networking sites. In: Olsen, Jr., D.R., Arthur, R.B., Hinckley, K., Morris, M.R., Hudson, S.E., Greenberg, S. (eds.) Proceedings of the 27th International Conference on Human Factors in Computing Systems, CHI 2009, 4–9 April 2009, Boston, MA, USA, pp. 201–210. ACM (2009)

19. Farooq, U., Kannampallil, T.G., Song, Y., Ganoe, C.H., Carroll, J.M., Giles, C.L.: Evaluating tagging behavior in social bookmarking systems: metrics and design heuristics. In: Gross, T., Inkpen, K. (eds.) Proceedings of the 2007 International ACM SIGGROUP Conference on Supporting Group Work, GROUP 2007, 4–7 November 2007, Sanibel Island, Florida, USA, pp. 351–360. ACM (2007)

20. Gupta, P., Goel, A., Lin, J., Sharma, A., Wang, D., Zadeh, R.: Wtf: the who to follow service at twitter. In: Schwabe, D., Almeida, V.A.F., Glaser, H., Baeza-Yates, R.A., Moon, S.B. (eds.) 22nd International World Wide Web Conference, WWW 2013, 13–17 May 2013, Rio de Janeiro, Brazil, pp. 505–514. International World Wide Web Conferences Steering Committee/ACM (2013)

21. Guy, I., Carmel, D.: Social recommender systems. In: Proceedings of the 20th International Conference on World Wide Web, WWW 2011 (Companion volume), pp. 283–284. ACM (2011)

22. Guy, I., Chen, L., Zhou, M.X.: Introduction to the special section on social recommender systems. ACM TIST **4**(1), 7 (2013)

23. Guy, I., Ronen, I., Wilcox, E.: Do you know?: recommending people to invite into your social network. In: Conati, C., Bauer, M., Oliver, N., Weld, D.S. (eds.) Proceedings of the 2009 International Conference on Intelligent User Interfaces, 8–11 February 2009, Sanibel Island, Florida, USA, pp. 77–86. ACM (2009)

24. Hannon, J., Bennett, M., Smyth, B.: Recommending twitter users to follow using content and collaborative filtering approaches. In: Amatriain, X., Torrens, M., Resnick, P., Zanker, M. (eds.) Proceedings of the 2010 ACM Conference on Recommender Systems, RecSys 2010, 26–30 September 2010, Barcelona, Spain, pp. 199–206. ACM (2010)

25. Herlocker, J.L., Konstan, J.A., Borchers, A., Riedl, J.: An algorithmic framework for performing collaborative filtering. In: SIGIR 1999: Proceedings of the 22nd Annual International ACM SIGIR Conference on Research and Development in Information Retrieval, 15–19 August 1999, Berkeley, CA, USA, pp. 230–237. ACM (1999)

26. Liben-Nowell, D., Kleinberg, J.M.: The link prediction problem for social networks. In: Proceedings of the 2003 ACM CIKM International Conference on Information and Knowledge Management, 2–8 November 2003, New Orleans, Louisiana, USA, pp. 556–559. ACM (2003)

27. Manca, M., Boratto, L., Carta, S.: Design and architecture of a friend recommender system in the social bookmarking domain. In: Proceedings of the Science and Information Conference 2014, pp. 838–842 (2014)

28. Manca, M., Boratto, L., Carta, S.: Mining user behavior in a social bookmarking system - A delicious friend recommender system. In: Helfert, M., Holzinger, A., Belo, O., Francalanci, C. (eds.) DATA 2014 - Proceedings of 3rd International Conference on Data Management Technologies and Applications, Vienna, Austria, 29–31 August, 2014. pp. 331–338. SciTePress (2014)

29. Marlow, C., Naaman, M., Boyd, D., Davis, M.: Ht06, tagging paper, taxonomy, flickr, academic article, to read. In: Proceedings of the Seventeenth Conference on Hypertext and Hypermedia, HYPERTEXT 2006, pp. 31–40. ACM, New York (2006). http://doi.acm.org/10.1145/1149941.1149949

30. Mobasher, B., Cooley, R., Srivastava, J.: Automatic personalization based on web usage mining. Commun. ACM **43**(8), 142–151 (2000). http://doi.acm.org/10.1145/345124.345169

31. Pearson, K.: Mathematical contributions to the theory of evolution. iii. Regression, heredity and panmixia, Philosophical transactions of the royal society of London. In: Series A, Containing Papers of a Math. or Phys. Character (1896–1934), vol. 187, pp. 253–318, January 1896

32. Quercia, D., Capra, L.: Friendsensing: recommending friends using mobile phones. In: Bergman, L.D., Tuzhilin, A., Burke, R.D., Felfernig, A., Schmidt-Thieme, L. (eds.) Proceedings of the 2009 ACM Conference on Recommender Systems, RecSys 2009, 23–25 October 2009, pp. 273–276. ACM, New York (2009)

33. Ratiu, F.: Facebook: People you may know, May 2008. https://blog.facebook.com/blog.php?post=15610312130

34. Ricci, F., Rokach, L., Shapira, B.: Introduction to recommender systems handbook. In: Ricci, F., Rokach, L., Shapira, B., Kantor, P.B. (eds.) Recommender Systems Handbook, pp. 1–35. Springer, USA (2011)

35. Simon, H.A.: Designing organizations for an information rich world. In: Greenberger, M. (ed.) Computers, communications, and the public interest, pp. 37–72. Johns Hopkins Press, Baltimore (1971)

36. Xiong, H., Shekhar, S., Tan, P.N., Kumar, V.: Exploiting a support-based upper bound of pearson's correlation coefficient for efficiently identifying strongly correlated pairs. In: Proceedings of the Tenth ACM SIGKDD International Conference on Knowledge Discovery and Data Mining, KDD 2004, pp. 334–343. ACM, New York (2004). http://doi.acm.org/10.1145/1014052.1014090

37. Zhou, T.C., Ma, H., Lyu, M.R., King, I.: Userrec: a user recommendation framework in social tagging systems. In: Fox, M., Poole, D. (eds.) Proceedings of the Twenty-Fourth AAAI Conference on Artificial Intelligence, AAAI 2010, 11–15 July 2010, Atlanta, Georgia, USA. AAAI Press (2010)

Developing a Pedagogical Cybersecurity Ontology

Soon Ae Chun[1] and James Geller[2(✉)]

[1] City University of New York – College of Staten Island,
Staten Island, NY, USA
Soon.Chun@csi.cuny.edu
[2] New Jersey Institute of Technology, Newark, NJ, USA
james.geller@njit.edu

Abstract. We present work on a hybrid method for developing a *pedagogical cybersecurity ontology* augmented with teaching and learning-related knowledge in addition to the domain content knowledge. The intended use of this ontology is to support students in the process of learning. The general methodology for developing the *pedagogical cybersecurity ontology* combines the semi-automatic classification and acquisition of domain content knowledge with pedagogical knowledge. The hybrid development method involves the use of a seed ontology, an electronically readable textbook with a back-of-the-book index, semi-automatic steps based on pattern matching, and the cyberSecurity Ontology eXpert tool (SOX) for an expert to fill the knowledge gaps. Pedagogical knowledge elements include importance, difficulty, prerequisites and likely misunderstandings. The pedagogical cybersecurity ontology can be more useful for students than an ontology that contains only domain content knowledge.

Keywords: Pedagogical cybersecurity ontology · Hybrid cybersecurity ontology development · Security ontology expert tool · Security domain knowledge · Textbook index · Augmented security ontology

1 Introduction

According to Guarino et al. [1] *"Computational ontologies are a means to formally model the structure of a system, i.e., the relevant entities and relations that emerge from its observation, and which are useful to our purposes."* This definition leaves open the possibility that the same *system* may be described in different ways to be *useful* to different communities of users. Thus, an ontology could be useful as a knowledge representation for a set of communicating application programs. In this article, we are interested in the two interconnected goals of creating an ontology for domain knowledge and tailoring it to the needs of a community of learners. Thus, the ontology should be useful as part of an educational environment. An ontology that incorporates both domain content knowledge and pedagogical knowledge can be described as containing *pedagogical content knowledge,* using the term of Shulman [3, 4]. In this work, we are constructing a cybersecurity ontology augmented with pedagogical knowledge, which we call *pedagogical cybersecurity ontology*, building on our previous work [2].

© Springer International Publishing Switzerland 2015
M. Helfert et al. (Eds.): DATA 2014, CCIS 178, pp. 117–135, 2015.
DOI: 10.1007/978-3-319-25936-9_8

1.1   Hybrid Method for Ontology Construction

Building ontologies is a difficult task. Two major kinds of approaches have defined the state-of-the-art. One approach relies primarily on human experts. This works well as long as the desired ontology is small. However, most useful ontologies are large. For example, the Standardized Nomenclature of Medicine – Clinical Terms (SNOMED CT) [5] contains on the order of 400,000 concepts. Due to differing viewpoints of human experts, large ontologies cannot be built in a completely modular fashion. Rather, experts working on different ontology modules need to communicate with each other. In many cases they need to negotiate an agreement on certain concepts, which increases the amount of time that is necessary for building the ontology.

The alternative approach is to automate the process by letting a computer read web pages, books, etc. just like a human would read them [6]. Unfortunately, perfecting this approach requires solving the natural language understanding task at the level of the Turing Test. This has not been achieved yet.

This leads us to the following hybrid approach. (1) Use any relevant structured or semi-structured information about the domain that is available, in addition to free text. (2) Implement a software tool that maximizes support for human experts and minimizes the time they need to spend on ontology building.

An et al. [7] and Geller et al. [8] have previously used the Deep Web as a source of structured information. However, this approach is inhibited by many web site owners. Thus, we have turned to another source of semi-structured domain knowledge, namely the alphabetic indexes of textbooks. The idea of an "alphabetic index" is old; Cleveland & Cleveland [9] mention an example from the 5th century. Automating the process of indexing a book has become a subject of Artificial Intelligence [10]. Our approach is to "mine" the knowledge that went into building an index and of the structure of the index as it is intertwined with the text in the book.

We have used a hybrid approach to developing a cybersecurity ontology. First, using an existing security ontology as a seed ontology, we have developed an auto-mated bootstrapping algorithm to extend it by adding more domain concepts [12]. The resulting security ontology is then extended by a domain expert using a knowledge acquisition tool. For this purpose, we developed the Security Knowledge Acquisition Tool (SKAT) for security domain experts to further refine the ontology [2].

In this paper, we are describing a second generation tool, called the *Security Ontology eXpert tool* (SOX) that was developed to support ontology building. While we focus on the application domain of cybersecurity for a college class, our approach, including the theory and tool, is generally applicable for any course topic domain.

The semi-automatic ontology construction component of the hybrid approach uses the textbook index as input and classifies each index term into a seed ontology of concepts.[1] [2, 11, 12]. However, this semi-automatic processing did not solve the classification problem, leaving many book index terms unclassified in the ontology. The SOX tool allows a domain expert to manually place these unclassified terms as concepts into the ontology. S/he may also enter new "bridge concepts" that are not in

---

[1] This "classification" is not related to the Description Logic classification algorithm.

the book index into the ontology to restructure the semi-automatically constructed ontology. These flexible functionalities allow the domain experts to edit the concepts as well as the structure to refine the ontology.

In order to reduce the cognitive burden on the domain expert, SOX parses the textbook to identify index terms co-occurring with ontology concepts to make suggestions of candidate concepts and relationships. Thus, SOX incorporates a Concept Recommendation component to identify occurrences of ontology concepts in the unstructured text of a textbook that co-occur with the semi-structured index terms, and it recommends these index terms as candidate concepts to help the expert user to identify the related concepts. This will minimize the expert's effort to search in the ontology structure for concepts that may be related to a security term from the textbook. The SOX tool contains a number of significant improvements over the previous generation tool [2].

1.2 Pedagogical Content Knowledge

A teacher needs to have subject matter knowledge, otherwise he cannot teach. However, many excellent subject matter experts are not good teachers, indicating that more is needed than just knowing what to teach. This is not a new observation. A teacher needs to know *what* to teach and *how* to teach it. It would be convenient to look at these two repositories of knowledge as "orthogonal" or as "independent modules." Thus, a teacher who has acquired the "how to teach module" could later "plug in" a new "what to teach module" and be ready to step in front of the students of a class s/he has never taught before. However, since at least 1986, when Lee Shulman wrote his paradigm-defining paper on pedagogical content knowledge, it is generally accepted that this modular framework does not correctly reflect the teaching process.

After Shulman's [3, 4] introduction of pedagogical content knowledge, the notion was further elaborated by Rowan et al. [13]. Quoting from Rowan:

> ...pedagogical content knowledge ... is different from, teachers' subject matter knowledge or knowledge of general principles of pedagogy. ... pedagogical content knowledge is a form of practical knowledge that is used by teachers to guide their actions in highly contextualized classroom settings. ... In Shulman's view, this form of practical knowledge entails, among other things: (a) knowledge of how to structure and represent academic content for direct teaching to students; (b) knowledge of the common conceptions, **misconceptions**, and **difficulties** that students encounter when learning particular content; and (c) knowledge of the specific teaching strategies ... to address students' learning needs in particular classroom circumstances.[2]

One issue in developing an ontology to support an educational environment is representing the additional (pedagogical) knowledge of the teacher; but it is also necessary to address the *secondary information needs* of the students. Students have information needs for items that are customarily not included in any written course material, and are only made available orally and/or on request. Questions expressing secondary information needs cannot be answered with domain content knowledge.

---

[2] Bold font added by us.

Many of these secondary information needs are connected to attempts of the students to minimize their study efforts. For example, a common question addressed to teachers is "Will this be on the midterm exam"? This question really encompasses three different possible secondary information needs. The student might wonder what the assessment mode will be for this topic, an exam, homework or a project. Secondly, the student might wonder about timing. It might be obvious that this material will be assessed by an exam, but will it be on the midterm exam or the final exam? Thirdly, the student might conclude from the topic itself or from the manner of the instructor while teaching it that this material is tangential to the course or beyond the scope of the semester and might therefore not appear on any assessment. Therefore the question ("Will this be ...") is attempting to satisfy a secondary information need, not connected to primary information needs about the course material.

The knowledge required to address a secondary information need is naturally available to the instructor. However, when trying to incorporate it into an ontology, one must note that this knowledge is qualitatively different from the knowledge about the domain. It is also different from general teaching knowledge. Thus, a second layer of knowledge is required for satisfying secondary information needs in an educational setting. The SOX tool provides the functionality to add this kind of pedagogical knowledge into the cybersecurity ontology.

The remainder of this paper is organized as follows. We present related work in Sect. 2, and briefly review our concept classification approach and its results in Sect. 3. Section 4 presents our approach to augmenting an ontology further through the use of concept recommendation and the SOX ontology development tool, as utilized by a human expert. In Sect. 5, we present the pedagogical knowledge representation. In Sect. 6, we discuss the results and experience with the SOX tool as well as the limitations. Section 7 concludes with a summary and future research tasks.

2 Related Work

2.1 Cybersecurity Ontologies

Ontology development approaches can be divided into the purely manual, the purely automatic, and various hybrid methods. In the manual approach, a team of human experts collects domain concepts, organizes them into a subconcept (or IS-A) hierarchy and then includes additional information, such as semantic relationships between pairs of concepts or details about concepts. LoLaLi is an example of an ontology that was manually built [14]. Building LoLaLi manually was very time consuming.

The automatic methods attempt to build an ontology by parsing of English text. There are several variants for this approach, such as clustering, linguistic pattern matching, formal concept analysis, or ontology alignment. Hindle's work is based on the clustering approach [15]. Hearst et al. used a linguistic pattern matching approach to find semantic relationships between terms from large corpora [16]. Formal concept analysis was used for extracting monotonic inheritance relations from unstructured data [17, 18]. Another method for developing a comprehensive ontology is by ontology alignment. BLOOMS is an example for building an ontology by alignment or ontology matching from smaller ontologies [19].

In previous research, we used a methodology for automatic construction of a domain ontology, by combining WordNet [20] concepts with domain-specific concept information extracted from the web [7]. Methods based on unstructured data from the web suffer from web pages that may be changing rapidly. Pattanasri et al. [21] developed a textbook ontology using the index and the table of contents. The concepts in each ontology are cross-referenced with page numbers to refer to corresponding textbook segments or slide page numbers. Because our work with SOX focuses on a Cyber Security Ontology, we note that work on several preexisting security-related ontologies has been reported in the literature [22–28].

None of the cited ontologies contains pedagogical content knowledge. This is a gap that we are addressing in our research.

2.2 Pedagogical Ontologies

Research has been published on ontologies that have pedagogy itself as the content. Thus, Wang [29] describes what he calls "Ontology of Learning Objects Repository." This ontology ties together the "ontological categories" (classes) of learning subject, learning objective, delivery instrument, instructional method, assessment instrument and assessment outcome by semantic relationships such as Is_assessed_by, Is_achieved_through, etc.

However, this ontology is not sufficient for representing pedagogical content knowledge in an ontology. This is because of the "highly contextual classroom setting" of Shulman, cited above. For example, in (b) (Sect. 1.2 above, italicized quote) the "common misconceptions and difficulties … when learning a particular content" are mentioned. These misconceptions cannot be formalized in a topic-independent manner. They are tightly tied into the actual domain content itself, and therefore a separate representation of pedagogical knowledge for every subject matter is required, besides the domain ontology. In Shulman's own words [3]:

> We are gathering an ever-growing body of knowledge about the misconceptions of students and about the instructional conditions necessary to overcome and transform those initial conceptions.

3 Automatic Cybersecurity Ontology Development

In this section, we are describing previous stages of this project. Our previous work [2, 11, 12] described the bootstrapping approach to enrich a seed ontology using an ensemble of different algorithms to classify book index terms into the seed ontology. The bootstrapping approach started with the ontology of Herzog [25] as seed ontology. Terms from the book index of Goodrich & Tamassia [30] are extracted and classified under the existing classes in the seed ontology by assembling different matching algorithms and evidence boosting algorithms using sources such as WordNet [20].

We started with exact matching, followed by matching using a stemmer and incorporating subterms recognizable in the index by indentation. For example, "vulnerabilities" in the textbook index matches with the concept "vulnerability" after applying the stemmer. In the next step we used substring matching together with Wikipedia categories to place index terms into the seed ontology. For instance, "replay attacks" overlaps with the Wikipedia category "Cryptographic Attacks" under "attacks," thus the system concludes that the index term defines a subcategory of "attacks." Next, section and subsection headings as well as linguistic heuristics (e.g., in a noun-noun phrase the second noun often indicates a superclass), etc. are used. For instance, "cryptographic compression function" belongs to "Cryptography" as a section heading and as a security class name. Next, prefix and postfix modifier matching is used.

In addition, NIST's security term definitions were extracted and included in the ontology to define its concepts [31]. However, a sizable number of index terms remained unclassified. Among 724 index terms, 263 terms were successfully classified into the seed ontology, which corresponds to 36.32 %.

In the next step, we are approaching the problems of including concepts and incorporating semantic relationships between the remaining unclassified index terms and existing concepts in the ontology. It is important to note that by including an index term in the ontology with an IS-A link, the term is promoted to a concept. Thus, we can talk about relationships between pairs of concepts. While including semantic relationships, we confront the following problem.

Assuming that there are N concepts in an ontology, the formula for the number of distinct pairs of concepts is $(N^2 - N)/2$. For a moderately sized ontology of 1000 concepts, that would mean 499,500 possible pairs. If a domain ontology allows for the use of IS-A relationships and nine other semantic relationships between pairs of concepts, an expert would need to consider each one of them for every pair of concepts. In reality, ten is a gross underestimate. If an expert could make a decision about each pair in ten seconds, he would need 173 work days of 8 h to review all pairs. Resources at this level are rarely available, as experts are normally busy in their field of expertise. Thus, the task of assigning semantic relationships must be minimized as much as possible by presenting only pairs to the expert that are highly likely to have a useful relationship. The SOX tool implements this idea.

4   SOX: Cybersecurity Ontology Expert Tool

4.1   Basic Assumptions

Our work is based on the following four heuristics [2].

H1. If a word or a multi-word term appears in the index of a book, then it describes an important domain concept for the ontology of this domain.

H2. If two index terms appear close to each other in the body of a book, then it is likely that there is a subclass relationship or a semantic relationship between them.

H3. If two index terms appear repeatedly close to each other, then the likelihood of a relationship between them is higher and/or the importance of the relationship is higher than for one appearance.

H4. In a well written textbook, sentences are semantic units. Thus, the basic unit of being "close to each other" will be the sentence, as opposed to a k-word neighborhood. Of course, other possible units also exist, such as paragraphs.

4.2 Formal Definition of Cybersecurity Knowledge Structure

The knowledge structure that is the backbone of SOX will be formally introduced in this section [2].

Definition 1. A sentence S is a sequence of words forming a grammatically correct English sentence terminated by a period.

Definition 2. A book image B of a book is the set of all sentences derived from the text of a book by removing front matter, back matter, figures, tables, captions, page headers, page footers, footnotes and all levels of chapter and section headers.

A book index is an alphabetical list of domain terms, together with the numbers of pages where the terms appear, possibly with a multi-level structure, synonyms, etc. However, an index will be viewed as just a list of terms in this formalism.

Definition 3. An index I is a one-level list of unique index terms T_i.

$$I = \langle T_1, T_2, \ldots T_m \rangle \tag{1}$$

We assume the existence of a basic ontology. (Otherwise, it needs to be created.)

Definition 4. The concept list C of an existing basic ontology consists of a one-level list of all the concepts C_i of the ontology.

$$C = \langle C_1, C_2, \ldots C_r \rangle \tag{2}$$

Furthermore, a concept in C is represented by its preferred English term, as opposed to an ID.

Definition 5. A term-concept co-occurrence pair TCP_{ij} is an ordered pair that consists of a term T_i from I (that is not in C) and a concept C_j from C such that there exists a sentence S in B that contains both T_i and C_j identifiable among the words of S.

$$TCP_{ij} = \langle T_i, C_j \rangle \in S \ \& \ S \in B \tag{3}$$

Definition 6. The ranked list RL of term-concept pairs is the list of all TCPs

$$RL = \langle \langle T_i, C_j \rangle \ldots \langle T_i, C_k \rangle \ldots \langle T_m, C_n \rangle \ldots \langle T_m, C_p \rangle \rangle \tag{4}$$

where if $F(<T_i, C_j>) \geq F(<T_m, C_p>)$, then $<T_i, C_j> \ll <T_m, C_p>$ where $F(.)$ represents the frequency ordering function of the co-occurrences between the term and concept and $\ll$ represents the *precede* relation in RL. The frequency ordering function $F(.)$ of coocurrence pairs is derived from the following two frequency functions for each component in the pairs.

Using $f(.)$ as the function that returns the frequency of terms or concepts mentioned in RL and the symbol $\ll$ to indicate "precede" then this would be expressed as

$$<T_i, C_j> \quad \ll \quad <T_m, C_n> \quad \leftrightarrow \quad f(T_i) \geq f(T_m)$$

Furthermore, if the frequency of $T_i = T_m$ then the same condition holds for the C_i. In other words, if $<T_i, C_j>$ appears before $<T_i, C_k>$ then the frequency of C_j within pairs occurring within sentences is greater than or equal to the frequency of C_k. We repeat that $f(.)$ is based on the frequencies in B, not on RL, because pairs in RL are unique.

$$<T_i, C_j> \quad \ll \quad <T_i, C_k> \quad \leftrightarrow \quad f(C_j) \geq f(C_k)$$

Definition 7. The projected term list PT consists of all unique terms in the order imposed by RL.

$$PT = <T_1, \ldots T_i, T_j, \ldots T_l>$$

such that $T_i \ll T_j$ in RL for all i and j.

Definition 8. The projected concept list of a term T, PC(T), consists of all unique concepts in the order imposed by RL that appear with T in a term concept pair.

$$PC(T) = <C_1, \ldots C_i, C_j, \ldots C_l>$$

such that $<T, C_i> \ll <T, C_j>$ for all i and j and all C_x appearing in pairs $<T, C_x>$ are included in PC(T). PC(T) will be used below to specify precisely what a user of the SOX tool sees on the screen. This formalism makes the optimistic assumption that there will be a TCP available in B for every term from I. This is not the case. In previous work, we described a transitive linkage mechanism to improve this situation [2].

5   Representing Pedagogical Knowledge

5.1   Pedagogical Knowledge of Misconceptions

We will focus on the issue of "misconceptions" as an example of developing pedagogical knowledge for an ontology. We argue that even though the most common misconceptions will be different in every content area, and therefore the relevant pedagogical knowledge will not be transferable between content areas, the representation of the *teacher's knowledge that some content-related concepts is susceptible to a misconception* based on his/her experience is possible in a generic way in every content area. It makes the task of building an ontology representing pedagogical content knowledge (PCK), i.e., a combination of content and pedagogical knowledge [3, 4], more manageable by providing standard building blocks for this purpose.

The proposed generic building blocks are demonstrated by two examples of misconceptions, one from K-12 (kindergarten to high school) education and another one from Computer Science. At the K-12 level, a student might have seen pictures or videos of whales, and might have concluded without any instruction or outside information that a whale is a fish, because it looks and swims like a fish. Ironically, a killer whale is not even a whale but belongs to the family of dolphins (Delphinidae), a fact unknown to this author until researching this article, which indicates that simple categorical misconceptions are not limited to K-12 students.

As an example of a common misconception in Computer Science education at the college level, we note that many students who have taken their first programming course appear to have the vague idea that any problem that can be expressed in a formal language and entered into computer memory is solvable by the computer. In appropriate theory classes they are then disabused of this misconception and confronted with the fact that there are even two kinds of "unsolvable" problems. A problem may be in principle not solvable with an algorithm (e.g., the Halting Problem) or solving it may take intractable amounts of time for a "large" problem instance (NP-complete problems). We will now review these two misunderstandings from an ontology perspective.

In the Computer Science study of ontologies (as opposed to philosophical ontology, see Guarino et al.'s definition [1]), the most common ontology format consists of concepts that are interconnected by IS-A links, which express category assignments of the form "Whale IS-A Mammal" as the backbone of all represented knowledge. We note that researchers have introduced fine distinctions for ontologies themselves (e.g., distinguishing between ontologies, terminologies, thesauri, controlled vocabularies, etc.) and between ontology elements (e.g., distinguishing between classes and concepts, and between A-Kind-Of relationships, IS-A relationships, subclass relationships, etc.).

Some researchers even distinguish between relations and relationships [32]. Thus, Bodenreider et al. [32] use "…relationship to refer to the links among concepts in ontologies (e.g., location_of). In contrast, [they] use relation to refer to the association between two concepts linked by some relationship." This cornucopia of meta-ontological terminology is further complicated by synonymous descriptions of ontology elements, such as lateral relationships and semantic relationships, which are widely considered to have the same meaning, as well as the usually synonymous terms "relationship" and "link." We will always talk about ontologies and concepts, but we interchangeably use IS-A and "subclass," "lateral relationship" and "semantic relationship" as well as "relationship" and "link."

In the ontology literature, graphical representations are commonly used to elucidate specific notions. In such a graphical representation, concepts are shown as boxes or ovals. Relationships are shown as arrows, annotated by names. The IS-A relationship is sometimes shown by a special kind of arrow, e.g., a bold arrow or a double-line arrow, omitting the label "IS-A." Thus the fact that a whale IS-A mammal may be shown as in Fig. 1, while the misconception for the whale concept is that the student usually has a concept for "fish," a concept for "mammal" and a concept for "whale" and the connections between those concepts are wrong as shown in Fig. 2.

For the computability concept in Computer Science, the student is most likely missing the concept of a "computationally unsolvable problem" or possibly the student has two concepts "solvable problem" and "computationally representable problem" and a wrong IS-A link from the latter to the former. Quite likely a student who has had some initial success in mastering programming will have these representational deficiencies, a missing concept and a wrong IS-A relationship. However, we did not work with human subjects to confirm this assumption.

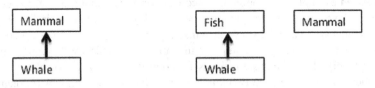

Fig. 1. Concept relationship "Whale IS-A Mammal".

Fig. 2. The misconception that a "Whale IS-A Fish".

Thus a teacher needs to perform "knowledge engineering" on her/his students. In the simpler case of the whale, one IS-A link will need to be deleted and another IS-A link will need to be created. As real deletion in the mind of the students is of course impossible, the IS-A link from whale to fish will need to be annotated as invalid.

But how does the knowledge structure in an ontology augmented with pedagogical knowledge need to look like in order to enable such operations? Or one might wonder what the knowledge structure of the teacher would look like. We propose to include the correct "whale IS-A mammal" node-link-node structure and a new relationship "misconceived-IS-A" from whale to fish in the augmented ontology (Fig. 3).

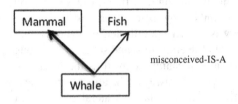

Fig. 3. Representing a categorical misconception in an ontology.

While every content domain needs to be analyzed where to place "misconceived-IS-A" links, the link itself can be used universally in every content domain.

5.2   Other Pedagogical Knowledge Representation in Ontologies

In this section, we discuss incorporating other pedagogical knowledge such as "importance," "difficulty," and "prerequisite-of" in the ontology [2].

- The "difficulty" attribute: This attribute provides information about how difficult it is to understand a concept in a domain.
- The "importance" attribute: This attribute provides information about how important an existing concept is toward the understanding of a domain.
- The "prerequisite-of" relationship: This relationship connects two concepts, such that the first concept is a prerequisite for understanding the second concept.

We note that difficulty, importance and prerequisite-of are expressed specifically for content *concepts*. Each one of these pedagogical elements appears to be universal, i.e., content-domain independent. While these three augmentations might well apply to relationships also, this remains future work. Each of these pedagogical elements is potentially useful for a student to satisfy secondary information needs as follows:

1. A student who is informed that a concept is difficult will know to allocate more time to it, or to study it when s/he is well rested, or to study it when s/he has access to a "helper" such as a tutor or the professor.
2. A student who needs to study a set of concepts under time constraints will first study all the concepts that are marked as important and will then advance to the other concepts, if time allows.
3. A student who has identified a book page or a recorded lecture about a specific concept might have problems understanding the explanation of the concept. In that case, s/he will use the prerequisite-of links to locate and review all the concepts expected to be known before studying the new concept. S/he will then retry to understand the concept that initiated the search for prerequisites.

In addition, these pedagogical elements can be used by the teachers to personalize the content in terms of student-levels. Personalized teaching among a diverse student body in one classroom has been a constant challenge for many teachers, often forcing them to go midstream, resulting in boring lectures for the high-achievers, and yet not being understandable to students who do not know the basics of the topic.

5.3   Capturing the Pedagogical Knowledge in Cybersecurity Ontology

In our previous work [2], we made use of the complete text of a machine-readable cybersecurity textbook [30] that is used in support of a junior-level college class on computer security. We especially made use of the index in the back of the textbook. Terms from a second textbook index [33] were merged into the list of terms. Then we eliminated (by hand) terms from the textbook index that did not appear specifically relevant as computer security concepts, for example "Yahoo." Thus, we reduced a list of 809 terms to 724 terms. Figure 4 shows the distribution of cybersecurity terms according to one source. The 22 most common terms are shown in Table 1. All other terms appear fewer than 15 times.

The frequency data can be used to formulate a heuristic for the importance of a concept. A concept that appears many times in the textbook is presumably important. We also performed a co-occurrence analysis for pairs of index terms. According to this analysis, the index term "web server" appears in the textbook of Goodrich and Tamassia 164 times (with another index term in the same sentence). The term "session" appears 132 times, while the term "certificate" appears 86 times. As might be expected, there is a long tail of terms that appear only once. The advantage of using only terms from the book indexes is that no special effort needs to be made to exclude stop words and little effort was necessary to recognize terms that describe concepts in cybersecurity.

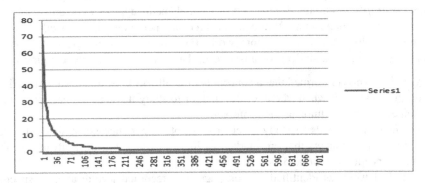

Fig. 4. Frequency of cybersecurity terms in one source (X-axis represents term IDs).

Finding a heuristic for prerequisite relationships is more difficult. We performed experiments with pairs of terms (A, B), where A appeared in the body of a section and B in the title of the section, assuming that understanding B required an understanding of A. However, the yield of these experiments was too low to be useful. Another possible heuristic could be based on two terms appearing repeatedly together in one section, however, only one of them appears before that section. No experiments were performed with this heuristic. Thus the determination of prerequisites will have to be made by a human expert in almost all cases.

Once prerequisites have been established they provide a heuristic for difficulty. Thus, a concept that has many prerequisites is more difficult to understand than a concept that does not. Some textbooks have a linear structure where every chapter builds on the previous chapter(s). In such a book the page number where an index term appears for the first time provides a weak heuristic of difficulty. Terms that are introduced later are likely to be more difficult. However, many textbooks do not have a linear structure. For example, the Elmasri & Navathe database textbook [34] defines in its preface several different paths through the material, including a data modeling-oriented path (high level) and a query language-oriented path. In such a textbook the page number of first appearance does not provide a useful heuristic for difficulty.

Table 1. Most common cybersecurity terms based on one source after cleaning.

TERM	FREQUENCY
Cryptography	71
Operating systems security	59
Physical security	58
Malicious software	49
Microsoft windows	43
Internet protocol	35
Advanced encryption standard	30
Linux	29
Transmission control protocol	28
Buffer overflow	27
Hypertext transfer protocol	25
Password	25
Domain name system	21
Authentication	20
Cryptography symmetric	20
Intrusion detection	19
Javascript	18
Operating systems concepts	17
Rivest-shamir-adleman cryptosystem	17
Application security	16
Ciphertext	16
Hypertext markup language	16

6 Pedagogical Cybersecurity Ontology Development Tool

6.1 A Tool for Acquiring Knowledge from a Textbook and from an Expert

The SOX (Security Ontology eXpert) tool was implemented to develop an augmented cybersecurity ontology for pedagogical use. The SOX tool uses information from the index and the book image of a cybersecurity textbook [30] to elicit the correct relationships between pairs of a term and a concept. SOX extends the functionality and speed of the previously developed SKAT tool [2] and corrects a number of problems of SKAT.

Figure 5 shows a screen dump of the main screen of the SOX tool. This screen is subdivided into several subwindows. The upper left subwindow presents the projected term list (Definition 7) to the domain expert, ordered by frequency value or in alphabetical order, as desired. The idea is that the user will start with the most commonly co-occurring terms, which, by the above heuristics, are most likely to take part in a semantic relationship of the domain.

Because the IS-A relationship is of paramount importance in ontologies, the domain expert is forced to first connect the chosen term from PT with an IS-A link to a concept in the existing ontology. The expert can either drag and drop the term into the ontology that appears in the lower right subwindow of the SOX interface; or s/he can select one of the concepts in the projected concept list PC(T) (Definition 8) for that term in the upper right subwindow.

To guide the expert concerning how to connect two concepts, the center of the main screen, below the box "Create new Relationship," displays all the sentences in which the ontology concept and the index term appear together. This is an important additional source of information that was not available in SKAT and is helpful to human experts.

Once the term from PT has been connected by an IS-A link to the ontology it is considered promoted to a concept, and the expert may continue with another term, or may assign one or more semantic relationships to the newly created concept. For this purpose s/he may select an already existing relationship from the menu in the upper middle of SOX or add a new relationship.

The relationship menu is initialized with a list of semantic relationships containing <IS-A, PART-OF, RELATED-TO, KIND-OF>.

SOX supports an undo and a redo button. Changes to the ontology are first saved into a database table and later reviewed by our project team before they are made permanent in the augmented ontology. This allows us to detect and reject any contaminations of the ontology due to accidental misuse of the tool by a domain expert. A special administrator login is implemented for this purpose.

6.2   Acquisition of Pedagogical Knowledge

One major feedback from the domain experts concerning the previous prototype system SKAT was that it combines the domain content knowledge and the pedagogical knowledge into one screen. In SOX, the two were separated and therefore importance and difficulty do not appear in Fig. 5. The left black side bar on the main screen shows

Fig. 5. SOX tool client.

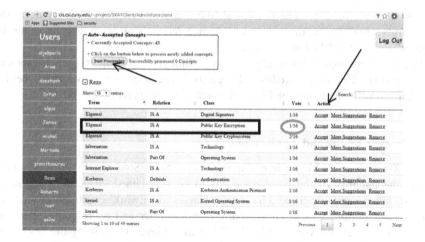

Fig. 6. SOX administrator view screen.

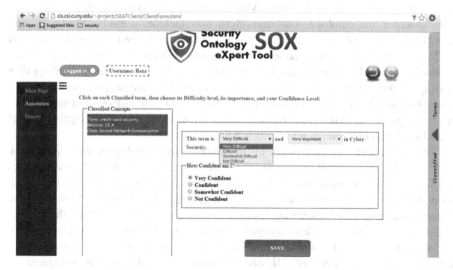

Fig. 7. Pedagogical knowledge input screen.

that there are two other screens of the tool, one for annotations with PCK and one for the history of interactions with the tool.

Domain experts need to log in, in order to have accountability for the decisions made. SOX records the input of each domain expert together with time stamps, in order to evaluate the time it takes the expert for every decision. SOX supports an Evaluation and a Production mode.

Figure 6 shows the administrator's console to view the concepts entered by a domain user in SOX. One of the suggested concepts "Elgamal" is categorized to be a sub-class of "Public Key Cryptosystem" by two users (out of 16), as seen in the "Vote" column. The administrator can process all the accepted categorizations by using the "Start Processing" button on top, or accept each relationship separately by using the "Accept" link in the "Action" column.

In Fig. 7, pedagogical knowledge, such as "difficulty" or "importance," is captured for the proposed concept "credit-card security." The user can also specify her/his confidence level concerning the IS-A relationship to the parent concept.

6.3   Current Status and Applications

The SOX tool is available at: http://isecurelab.info:8080/SoxClient/ with access control, as the integrity of the ontology needs to be maintained. SOX has been made available to a domain expert who has taught classes in network security to evaluate its usability and utility. SOX is substantially user-friendlier and its response times are faster than the previous version SKAT. This was achieved by limiting the amount of client-server traffic and doing more of the processing on the client side. We are also caching the ontology on the client side.

In many cases the domain expert noticed that there were concepts missing under which to include a new index term. Very fundamental concepts such as "software" were absent in the ontology. Therefore, SOX now includes a mechanism to add any new "bridge concept" on the fly under any other concept. If the new concept does not "fit" anywhere it needs to be added under the root concept *Thing*. Currently, we are evaluating the tool with the domain experts to expand the cybersecurity ontology. We also apply the cybersecurity ontology to allow students and teachers to search for concept-related public teaching materials, such as YouTube videos, Slidshare slides, academic papers in DBLP, as well as textbook pages [11]. This is an ongoing project available at http://isecurelab.info:8080/WASlob/.

The "SOX approach" can work for moderately sized ontologies when a textbook with an index is available. However, as mentioned above, many realistic ontologies have on the order of hundreds of thousands of concepts. For SNOMED CT [5] with about 400,000 concepts, *assuming no prior knowledge of relationships* between those concepts, the task is overwhelming. Using again the formula $(N^2 - N)/2$ for the number of distinct pairs, that would lead to approximately 79999800000 possible relationships, i.e., about 80 billion. This would require about 76,000 ontology experts to work a year without taking any days off. If one textbook with all concepts contained in SNOMED CT were available (which is not the case) the SOX approach would still be intractable. The task of building ontologies remains difficult.

7   Conclusions and Future Work

This paper presents both the theoretical/conceptual aspects and the tool implementation of our work on building a cybersecurity ontology to support education. We have extended the idea of augmented ontologies that contain both domain content knowledge and pedagogical knowledge. Specifically, we have added a first model for representing misunderstandings. On the tool side, we have created a cybersecurity ontology development tool (SOX) for domain experts. This tool allows the users to add domain content knowledge and pedagogical knowledge. Large scale testing of the effectiveness of the cybersecurity ontology in teaching is planned. The issue of different experts modeling the same concepts in different ways and how to harmonize such discrepancies is still a research problem. In addition to the difficulty and importance *of concepts* and the prerequisite-of relationship *between concepts,* complete coverage of pedagogical knowledge needs to be considered in the future.

Acknowledgements. This work is partially funded by NSF grants 1241687 and 1241976. We gratefully acknowledge Pearson and Addison-Wesley for making the electronic textbooks available. We acknowledge the contributions of Mickel Mansour for building the SOX prototype system and of the domain experts at NJIT, Reza Curtmola and Roberto Rubino, for their participation and their valuable feedback.

References

1. Guarino, N., Oberle, D., Staab, S.: What is an Ontology? In: Staab, S., Studer, R. (eds.) Handbook on Ontologies. International Handbooks on Information Systems. Springer, Heidelberg (2009)
2. Geller, J., Chun, S., Wali, A.: A hybrid approach to developing a cyber security ontology. In: Proceedings of the 3rd International Conference on Data management Technologies and Applications, Vienna, Austria, pp. 377–384 (2014)
3. Shulman, L.S.: Those who understand: knowledge growth in teaching. Educ. Researcher **15** (2), 4–31 (1986)
4. Shulman, L.S.: Knowledge and teaching: foundations of the new reform. Harvard Educ. Rev. **57**(1), 1–22 (1987)
5. Cornet, R., de Keizer, N.: Forty years of SNOMED: a literature review. BMC Med. Inform. Decis. Mak **8**(1), S2 (2008). doi: 1472-6947-8-S1-S2 [pii] 10.1186/1472-6947-8-S1-S2
6. Wijaya, D.T.: NELL: a machine that continuously reads, learns, and thinks by itself, May 2014. The Global Scientist, online magazine by international Fulbright science and technology fellows
7. An, Y.J., Geller, J., Wu, Y., Chun, S.A.: Automatic generation of ontology from the deep web. In: Proceedings Database and Expert Systems Applications, DEXA 2007, Regensburg, Germany (2007)
8. Geller, J., Chun, S.A., An, Y.J.: Toward the semantic deep web. IEEE Computer **41**, 95–97 (2008)
9. Cleveland, D.B., Cleveland, A.D.: Introduction to Indexing and Abstracting, Fourth edn. Libraries Unlimited Inc., Englewood (2013)
10. Wu, Z., Li, Z., Mitra, P., Giles, C.L.: Can back-of-the-book indexes be automatically created? In: Proceedings CIKM, San Francisco, CA, pp. 1745–1750 (2013)
11. Chun, S.A., Geller, J., Wali, A.: Developing cyber security Ontology and linked data of security knowledge network. In: Proceedings Conference of the Florida Artificial Intelligence Research Society (Flairs-27), Pensacola, FL (2014)
12. Wali, A., Chun, S.A., Geller, J.: A bootstrapping approach for developing cyber security ontology using textbook index terms. In: Proceedings International Conference on Availability, Reliability and Security (ARES 2013), University of Regensburg, Germany (2013)

13. Rowan, B., Schilling, S.G., Ball, D.L., Miller, R., et al.: Measuring teachers' pedagogical content knowledge in surveys: an exploratory study. Consortium for Policy Research in Education (2001). http://sii.soe.umich.edu/newsite/documents/pck%20final%20report%20revised%20BR100901.pdf. Accessed 12/December/2014

14. Caracciolo, C.: Designing and implementing an Ontology for logic and linguistics. Literary Linguist. Comput. **21**, 29–39 (2006)

15. Hindle, D.: Noun classification from predicate-argument structures. In: Proceedings 28th Annual Meeting of the Association for Computational Linguistics, Pittsburgh, Pennsylvania (1990)

16. Hearst, M.A.: Automatic acquisition of hyponyms from large text corpora. In: Proceedings 14th Conference on Computational linguistics, Nantes, France (1992)

17. Cimiano, P., Hotho, A., Staab, S.: Learning concept hierarchies from text corpora using formal concept analysis. J. Artif. Int. Res. **24**, 305–339 (2005)

18. Wiebke, P.: A set-theoretical approach for the induction of inheritance hierarchies. Electron. Notes Theor. Comput. Sci. **53**, 1–13 (2004)

19. Jain, P., Hitzler, P., Sheth, A.P., Verma, K., Yeh, P.Z.: Ontology alignment for linked open data. In: Patel-Schneider, P.F., Pan, Y., Hitzler, P., Mika, P., Zhang, L., Pan, J.Z., Horrocks, I., Glimm, B. (eds.) ISWC 2010, Part I. LNCS, vol. 6496, pp. 402–417. Springer, Heidelberg (2010)

20. Fellbaum, C.: WordNet: An Electronic Lexical Database. MIT Press, Cambridge (1998)

21. Pattanasri, N., Jatowt, A., Tanaka, K.: Context-aware search inside e-learning materials using textbook ontologies. In: Dong, G., Lin, X., Wang, W., Yang, Y., Yu, J.X. (eds.) Advances in Data and Web Management. LNCS, vol. 4505, pp. 658–669. Springer, Heidelberg (2007)

22. Souag, A., Salinesi, C., Comyn-Wattiau, I.: Ontologies for security requirements: a literature survey and classification. In: Bajec, M., Eder, J. (eds.) CAiSE Workshops 2012. LNBIP, vol. 112, pp. 61–69. Springer, Heidelberg (2012)

23. Fenz, S., Ekelhart, A.: Formalizing information security knowledge. In: Proceedings 4th International Symposium on Information, Computer, and Communications Security, Sydney, Australia (2009)

24. Geneiatakis, D., Lambrinoudakis, C.: An ontology description for SIP security flaws. Comput. Commun. **30**, 1367–1374 (2007)

25. Herzog, A., Shahmeri, N., Duma, C.: An ontology of information security. Int. J. Inf. Secur. Priv. **1**(4), 1–23 (2007)

26. Kim, A., Luo, J., Kang, M.: Security ontology for annotating resources. In: Meersman, R. (ed.) OTM 2005. LNCS, vol. 3761, pp. 1483–1499. Springer, Heidelberg (2005)

27. Undercoffer, J., Joshi, A., Pinkston, J.: Modeling computer attacks: an ontology for intrusion detection. In: Vigna, G., Kruegel, C., Jonsson, E. (eds.) RAID 2003. LNCS, vol. 2820, pp. 113–135. Springer, Heidelberg (2003)

28. Blanco, C., Lasheras, J., Valencia-Garcia, R., Fernandez-Medina, E., Toval, A., Piattini, M.: A systematic review and comparison of security ontologies. In: Proceedings of the Third International Conference on Availability, Reliability and Security (2008)

29. Wang, S., Ontology of Learning Objects Repository for Pedagogical Knowledge Sharing Interdisciplinary Journal of E-Learning and Learning Objects, vol. 4, Formerly the Interdisciplinary Journal of Knowledge and Learning Objects (2008)

30. Goodrich, M., Tamassia, R.: Introduction to Computer Security. Addison-Wesley, USA (2010)

31. Glossary of Key Information Security Terms. NIST Interagency Report, p. 222: NIST, US Department of Commerce (2012)

32. Vizenor, L.T., Bodenreider, O., McCray, A.T.: Auditing associative relations across two knowledge sources. J. Biomed. Inform. **42**(3), 426–439 (2009)
33. Skoudis, E., Liston, T.: Counter Hack Reloaded: A Step-by-Step Guide to Computer Attacks and Effective Defenses, 2nd edn. Prentice Hall PTR, Englewood Cliffs (2006). Paperback – January 2, ISBN-10: 013148104
34. Elmasri, R., Navathe, S.B.: Fundamentals of Database Systems, 6th edn. Pearson, Boston (2010)

A Reverse Engineering Process for Inferring Data Models from Spreadsheet-based Information Systems: An Automotive Industrial Experience

Domenico Amalfitano[1], Anna Rita Fasolino[1],
Porfirio Tramontana[1(✉)], Vincenzo De Simone[1],
Giancarlo Di Mare[2], and Stefano Scala[2]

[1] Department of Electrical Engineering and Information Technologies,
University of Naples Federico II,
Via Claudio 21, Naples, Italy
{domenico.amalfitano,anna.fasolino,porfirio.tramontana,
vincenzo.desimone2}@unina.it
[2] Fiat Chrysler Automobiles, Pomigliano Technical Center,
Via ex Aeroporto, Pomigliano d'Arco, Italy
{giancarlo.dimare,stefano.scala}@fcagroup.com

Abstract. Nowadays Spreadsheet-based Information Systems are widely used in industries to support different phases of their production processes. The intensive employment of Spreadsheets in industry is mainly due to their ease of use that allows the development of Information Systems even by not experienced programmers. The development of such systems is further aided by integrated scripting languages (e.g. Visual Basic for Applications, Libre Office Basic, JavaScript, etc.) that offer features for the implementation of Rapid Application Development processes. Although Spreadsheet-based Information Systems can be developed with a very short time to market, they are usually poorly documented or in some case not documented at all. As a consequence, they are very difficult to be comprehended, maintained or migrated towards other architectures, such as Database Oriented Information Systems or Web Applications. The abstraction of a data model from the source spreadsheet files represents a fundamental activity of the migration process towards different architectures. In our work we present an heuristic- based reverse engineering process for inferring a data model from an Excel based information system. The process is fully automatic and it is based on seven sequential steps. Both the applicability and the effectiveness of the proposed process have been assessed by an experiment we conducted in the automotive industrial context. The process was successfully used to obtain the UML class diagrams representing the conceptual data models of three different Spreadsheet-based Information Systems. The paper presents the results of the experiment and the lessons we learned from it.

© Springer International Publishing Switzerland 2015
M. Helfert et al. (Eds.): DATA 2014, CCIS 178, pp. 136–153, 2015.
DOI: 10.1007/978-3-319-25936-9_9

1 Introduction

Spreadsheets are interactive software applications designed for collecting and analyzing data in tabular form. In a spreadsheet, data are organized in worksheets, any of which is represented by a matrix of cells each containing either data or formulas. Modern spreadsheets applications (e.g. Microsoft Excel) are integrated with scripting languages (e.g., Microsoft Visual Basic for Applications). These languages allow the realization of interactive functionalities for the management of the underlying data by adopting Rapid Application Development processes. In these scenarios, spreadsheets have left behind their original role of calculation sheets, becoming instead core business tools. As a recent analysis showed, the usage of spreadsheets is diffused in a relevant community made by many millions of end-user programmers [32].

Although it is possible to build Spreadsheet-based Information Systems very quickly and without high development costs, their quality is often low. The data often present replication problems; data management tends to be error prone due to the lack of native consistency checking or data visibility restriction mechanisms. Therefore, the user dissatisfaction when working with these systems may increase as the size and the complexity of data increases, so business organizations may be compelled to look for more effective solutions.

Migrating a legacy Spreadsheet-based Information System towards new technologies and platforms often provides an effective solution to these problems. In the last years, several processes, techniques and tools to support the migration of legacy software systems have been proposed in the literature [9, 10, 22, 23].

The first and more critical step in a migration process consists in the reverse engineering of the data model of the information scattered in the cells of different spreadsheets and worksheets. The reverse engineered model will provide a useful abstraction about the analyzed data and can be considered the starting point for planning any reengineering activity.

In [6] we presented a process for migrating a legacy Spreadsheet-based Information System to a new Web application. This process was defined in an industrial context where we reengineered a legacy system used in a tool-chain adopted by an automotive company for the development of embedded systems. According to this company, such system was affected by many maintenance and usability issues, which motivated the migration process.

The first step of the migration process was devoted to the reverse engineering of a data model and of the business rules embedded in the spreadsheet system. The successive steps regarded the reengineering and reimplementation of the system according to the target platform. The first step was performed using a set of heuristic rules that we described in [7].

In [8] we presented a further refinement of our previous works: we proposed a structured reverse engineering process that can be used to analyze spreadsheet-based information systems and to infer a UML class diagram from them. The class diagram provides a conceptual model of the data embedded in the spreadsheets and is comprehensive of classes, class names, class attributes, association and composition relationships. The process is based on the execution of seven sequential steps that can be automatically performed. The effectiveness of the proposed process has been assessed by a wider experimentation involving more spreadsheet-based information systems from the same industrial domain.

At variance with our past work [8], in this paper we present a wider analysis of works related to our approach and highlight their main differences compared to ours. Moreover, we present the technologies and tools we adopted to carry out the reverse engineering process and describe the lessons we learned from our industrial experience.

The paper is organized as it follows: Sect. 2 presents related works, while Sect. 3 illustrates the proposed data model reverse engineering process. Section 4 illustrates the case studies we considered to evaluate the approach. Section 5 shows the lessons we learned while Sect. 6 presents conclusive remarks and future works.

2   Related Works

Due to the wide diffusion of spreadsheets in business and in industry, a great interest in spreadsheet analysis has been recently recorded. Several authors addressed this problem and proposed techniques and tools supporting the automatic analyses of spreadsheets.

Several works in the literature are aimed at inferring a data model from spreadsheets. Some of them are based on explicit extraction and transformation rules that need to be provided by the users, such as the technique proposed by Hung et al. [27].

The works of Abraham et al. [1–4] exploit the existence of shared templates underlying a given spreadsheet corpus, and propose techniques for automatically inferring these templates. The inferred template is hence exploited for safe editing of the spreadsheets by end-users, avoiding possible errors. Two other approaches with similar purposes are the ones of Mittermeir and Clermont [29] and Ahmad et al. [5] that exploit specific information about the type of data contained in the spreadsheet cells.

Cunha et al. [12] propose a technique to infer relational models from spreadsheets exploiting methods for finding functional dependencies between data cells. The same authors propose further techniques to infer ClassSheets Models from spreadsheets [13] by adopting the models proposed by Abraham and Erwig. They propose a tool for embedding the visual representation of ClassSheets Models into a spreadsheet system in [14]. The tool allows to work with the model ad its instance in the same environment at runtime. In following works, [15, 16], they propose a framework for the employment of a Model Driven Development approach in the spreadsheet context. The framework is able to support the bidirectional transformation between the model and the data of a spreadsheet. They evaluate the effectiveness of this approach in [17]. Moreover Cunha et al., in [18], propose a technique for the automatic generation of Class Diagram Models from ClassSheet Models. They use functional dependencies to improve spreadsheets. The inferred functional dependencies can be exploited to add visual objects on the spreadsheets to aid the users in data editing [19]. The same authors in [20] perform an empirical study in order to assess the impact of their Model Driven approach on users' productivity. In [21] they perform a further empirically study aimed at comparing the quality of automatically generated relational models against the ones provided by a group of database experts [21].

Chen et al. [11] propose automatic rules to infer some information useful in the migration of a spreadsheet into a relational database. They evaluated the effectiveness of the proposed rules on a very large set including more than 400 thousands of spreadsheets crawled from the Web.

The work by Hermans et al. [25] addresses a problem very similar to ours that is the recovery of a conceptual data model from spreadsheets. That paper indeed presents a technique for reverse engineering a data model from spreadsheets that is based on two-dimensional patterns. These patterns regard layout, data and formulas included in the spreadsheets and can be specified, recognized and transformed into class diagrams. Some patterns were proposed by the authors, as well as other ones were found in the literature (e.g. [28, 30, 31]). The technique has been validated with respect to the spreadsheets corpus proposed by Fisher and Rothermel [24]. Successively, Hermans et al. [26] propose an approach to abstract leveled dataflow diagrams from a spreadsheet to support its understanding and they evaluated the usefulness of this approach by involving end-users from a large financial company.

However, these techniques are mainly applicable to spreadsheets that are used as calculation sheets, and that perfectly follow the predefined patterns. On the contrary these techniques may not be effective for analyzing spreadsheets not conforming to that patterns or used as information systems. In the latter case, indeed, no formula is present in the sheets and the rules used by the authors to classify cells may be misleading. The spreadsheets developed in the considered industrial context were used as information systems too and were not calculation sheets. As a consequence, we were not able to reuse the techniques found in the literature as-are, but we had to adapt them to the specific context. In particular, we had to look for heuristic rules applicable to the considered types of spreadsheets and we had to define a process made of a sequence of steps for applying the selected rules. The process we defined will be illustrated in the following section.

3 The Conceptual Data Model Reverse Engineering Process

The industrial context of our work included a large number of spreadsheets, implemented by Excel files, used in the development process of Electronic Control Units (ECU) in an automotive company. They supported the Verification & Validation activities (V&V) and the management of Key Performance Indicators (KPI) about the development process. The spreadsheets inherited from a same template and included some VBA functionalities providing data entry assistance. Moreover, their cells followed well-defined formatting rules (i.e., font, colors, and size of cells) that improved the readability and usability of the spreadsheets, and complied to the following layout rule: all the data concerning the same topic in a single spreadsheet file were grouped together in rows or columns separated by empty cells or according to specific spreadsheets patterns [25].

Taking the cue from other works proposed in the literature, we founded our data model reverse engineering process on a set of heuristic rules. We defined a process made of seven steps that can be automatically performed in order to infer, with gradual refinements, the UML class diagram of the considered information system. In each step, one or more heuristic rules are executed. Each rule is based on the analysis of one or more spreadsheets belonging to the corpus of spreadsheet files composing the subject information system. In the following we describe the steps of the proposed process.

Step 1. In this step we preliminarily apply *Rule 1* that abstracts a class named *Sp* whose instances are the Excel files that comply with a same template.

Step 2. In this step we exploit the *Rule 2* that is executed on a single spreadsheet file of the corpus. This rule associates each non empty sheet of the Excel file with a class S_i, having the same name of the corresponding sheet. Moreover, an UML composition relationship between the *Sp* class and each S_i belonging to the file is inferred. The multiplicity of the association on each S_i side is equal to 1. Figure 1 shows an example of applying this rule on an example spreadsheet.

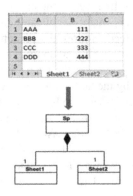

Fig. 1. Example of Step 2 execution.

Step 3. In this step we exploit the *Rule 3* that is executed on a single spreadsheet file of the corpus. This heuristic, according to [2, 25], associates a class A_j for each non empty cell area $Area_j$ of a sheet already associated to a class S_i by the *Rule 2*. Each class A_j is named as follows: *Name of the sheet_Area$_j$*. Moreover an UML composition relationship between the S_i class and each A_j class is inferred. The multiplicity of the association on each A_j side is equal to 1. Figure 2 shows an example of Step 3 execution.

Step 4. In this step two rules, *Rule 4.1* and *Rule 4.2*, are sequentially applied.

The *Rule 4.1* is applied to discriminate the header cells [1] of each area $Area_j$ that was inferred in the previous step. *Rule 4.1* analyzes the content of the spreadsheet files belonging to the corpus in order to find the invariant cells for each area $Area_j$. An invariant cell of an area is a cell whose formatting and content is the same in all the analyzed spreadsheets.

The set of invariant cells of an area $Area_j$ composes the header of that area. Figure 3 shows how the *Rule 4.1* works. In a first phase (Fig. 3-A) all the spreadsheets are queried to select, for each of them, the cells of a given area.

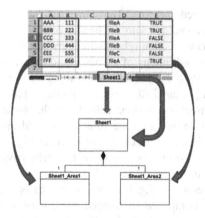

Fig. 2. Analysis of non-empty cell areas belonging to Sheet1 executed in Step 3.

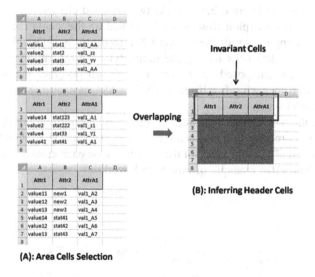

Fig. 3. Example of Step 3 execution.

In the next phase (Fig. 3-B) all the selected cells are analyzed to recognize the invariant cells for the considered area. Intuitively, if we imagine to overlap the contents of all the spreadsheets for a given area $Area_j$, then the header is given by the cells that, for the considered area, are invariant in all the files, as shown in Fig. 3-B.

Rule 4.2 is executed on the headers inferred by *Rule 4.1*. For each area $Area_j$, this heuristic analyzes the style and formatting properties of the cells composing its header. It permits to discriminate subareas $SubArea_m$ having header cells satisfying specific patterns. *Rule 4.2* associates a class SA_m for each $SubArea_m$. Each class SA_m is named as follows: *Name of the sheet_SubArea_m*, if no name could be associated to the class according to the recognized patterns. The names of the attributes of the class are inferred from the values contained into the header cells.

Moreover, an UML composition relationship between the class S_j and each related SA_m class is inferred. The multiplicity of the association on each class SA_m side is equal to 1. Some examples of *Rule 4.2* executions are reported in Fig. 4, 5, 6 and 7.

As an example, in Fig. 4 an UML class, having the default name *Sheet1_Area1*, is inferred since three consecutive header cells have the same formatting style. The attributes of the class are named after the values contained into the header cells.

Figure 5 shows a variant of the example reported in Fig. 4. In this case two UML classes are inferred since two groups of consecutive cells having the same formatting characteristics were found.

A further pattern is shown in Fig. 6 where the header is structured in two different levels. The upper level, composed of merged cells, permits to infer a class named *ClassA*, while its attributes are named after the values contained in the header cells of the lower level.

In the example shown in Fig. 7, the previous pattern is applied twice and a composition relationship is inferred between the two obtained classes.

Step 5. In this step we exploit Rule 5 that is applied on the whole corpus of spreadsheets. *Rule 5* is applied to all the subareas $SubArea_m$ to find possible sub-subareas made by groups of data cells having the same style and formatting properties. If a subarea $SubArea_m$ is composed by two or more sub-subareas then the heuristic associates a class, SSA_k for each sub-subarea. Each class SSA_k is named as follows: *Name of the sheet_SubSubArea_k*.

Moreover, the new classes substitute the class associated to the $SubArea_m$, and an UML composition relationship between the A_i class and each SSA_k class belonging to the area is inferred. The multiplicity of the association on each class SSA_k side is equal to 1. Figure 8 shows an example of how Step 5 works.

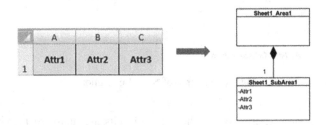

Fig. 4. Example of header cells pattern inferring a single class and its attributes.

Step 6. In this step the *Rule 6* is exploited. This heuristic is applied to the overall corpus of spreadsheets. It analyzes the formatting properties of the cells belonging to consecutive subareas and sub-subareas in order to infer association relationships and multiplicities between the classes that were associated to these areas in the previous steps.

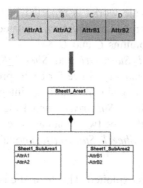

Fig. 5. Example of header cells pattern inferring two classes and their attributes.

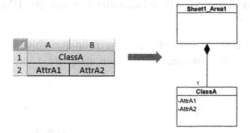

Fig. 6. Example of header cells pattern inferring a single class with attributes and class name.

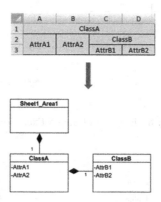

Fig. 7. Example of header cells pattern inferring two classes with attributes, class names and their composing relationship.

An example of how this rule works is reported in Fig. 9. In this case two consecutive subareas are considered, i.e., *SubArea1* related to columns A and B and *SubArea2* corresponding to columns *C* and *D*.

The classes named *Sheet1_SubArea1* and *Sheet1_SubArea2* were associated to *SubArea1* and *SubArea2*, respectively. Since for each spreadsheet a tuple of *SubArea1* is composed by a number of merged cells that is an integer multiple of the number of merged cells related to a tuple of *SubArea2*, then Rule 6 infers a UML association between the two classes related to the two subareas. The multiplicity of the association on the side related to the class *Sheet1_SubArea1* is equal to 1 whereas the one on the other side is 1.*

Step 7. In this step the *Rule 7* is exploited. This heuristic is applied to the overall corpus of spreadsheets. It analyzes the value of the cells in order to infer association relationships between classes. As an example, if in all the spreadsheets, for each cell of a column/row that was exploited to infer an attribute of a class *A* there is at least a cell of a column/row that was exploited to infer an attribute of a class *B* having the same value, then it is possible to define a UML relationship between the UML Class *A* and the UML Class *B*.

Moreover, if the cells of the column *A* have unique values then the *A* side and *B* side multiplicities of the inferred UML relationship are 1 and 1..* respectively, as shown in Fig. 10.

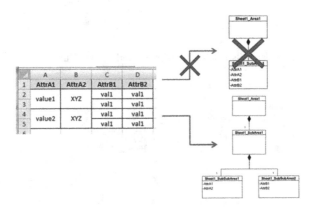

Fig. 8. Example of Step 5 Execution.

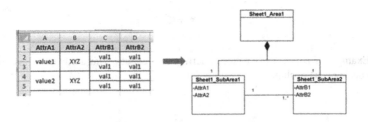

Fig. 9. Example of Step 6 execution.

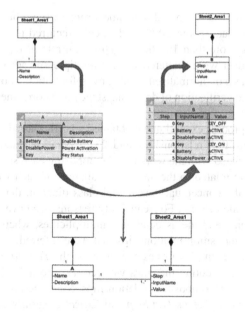

Fig. 10. Example of Step 7 execution.

4   Case Studies

To analyze the effectiveness of the proposed process, we performed a series of case studies involving different spreadsheet-based information systems from the same industrial domain. The goal of these case studies was to evaluate the applicability of the process and the acceptability and precision of the inferred models.

In the first case study we used the process to reverse engineer the conceptual data model from a legacy Spreadsheet-based Information System implemented in Microsoft Excel. This system was used by the HIL (*Hardware-in-the-Loop*) Validation Team of Fiat Group Automobiles and provided strategic support for the definition of high-level testing specifications, named *Test Patterns*. Test Patterns represent essential artifacts in the overall testing process [33] since they allow the automatic generation of test cases, the *Test Objects*, necessary to exercise the Electronic Control Units (*ECUs*) of automotive vehicles. The generation process is carried out thanks to a lookup table that embeds both the translation of high-level operations to the corresponding set of low-level instructions, and the mapping between input and output data to the set of ECU pins.

Test Patterns were implemented by means of a predefined Excel template file organized into 7 worksheets, referred to different phases of the testing process.

The worksheets embedded the model of the data. In the considered context, such template has been adopted to instantiate a set of 30,615 different Excel files, constituting the overall spreadsheet-based informative system. Each spreadsheet contained on average 2,700 data cells.

All of these files inevitably shared the same structure (i.e., the data model) with a high rate of replicated data (about 20 % of data cells recurred more than 1,300 times, while about the 50 % more than 100 times). This high rate of replication was mainly due to the fact that data were scattered among multiple spreadsheets. Therefore the underlying model was not normalized, with any information about related data. In addition, the automatic verification of data consistency and correctness was not natively supported.

At the end of the process execution we obtained a UML class diagram composed of 36 classes, 35 composition relationships and 23 association relationships. Seven of these classes are associated with the worksheets composing each spreadsheet file, while the remaining ones are related to the areas and subareas of each worksheet.

Figure 11 shows the conceptual data model class diagram that was automatically inferred by executing the process. For readability reasons, we have not reported all the association relationships and the associations' multiplicities, whereas we reported all the classes and the composition relationships that were inferred.

Figure 12 shows an example of rules execution on the *Test Pattern* sheet that is the most complex one. By executing the rules we were able to infer five classes, 4 association relationships and 5 composition relationships. Moreover, it shows how the *Rule 4* was able to infer the two classes *Test Step* and *Expected Results* and *Rule 6* proposed the association relationship between them.

In the second and third case study, we analyzed two further systems used by the company. The systems included a Software Factory Test Case (called SW TC) repository and a KPI repository (hereafter KPI).

The SW TC repository contains essential artifacts of MIL (*Model-in-the-Loop*) and SIL (*Software-in-the-Loop*) testing process in the considered company, since they allow the automatic generation of executable MATLAB test scripts necessary to

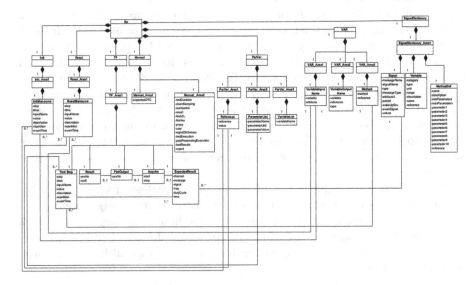

Fig. 11. Inferred conceptual UML class diagram for the Test Pattern Information System.

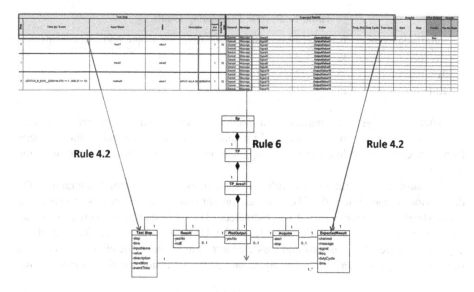

Fig. 12. Example of class proposal and its mapping with the spreadsheet file.

validate both the models and the source code of the software components that will be deployed on the ECUs. This information system is composed by 14,000 Excel files inheriting from a common template composed by 10 sheets. Each spreadsheet contained 4,000 data cells on average. About the 75 % of data cells are replicated in the spreadsheets.

KPI repository is a spreadsheet-based information system used to manage the key performance indicators of each development project regarding a software component belonging to a specific type of vehicle. The information system is composed by 1,500 Excel files inheriting from a common template composed by 8 sheets. These spreadsheets contained 1,370 data cells on average. About the 77 % of data cells are replicated in the spreadsheets.

To support the process execution a prototype software application was developed. It is implemented in C# and takes as input the location of the spreadsheets composing the information system under analysis. It accesses to the underlying Excel object model[1] of each spreadsheet in order to extract the raw data that are needed for the automatic execution of the steps, i.e., spreadsheet's structure, data content, properties of cells, etc. The tool is able to provide as output an XMI file representing the inferred Class Diagram. To implement the Step 7 the prototype exploits some features offered by the Data Quality tool Talend Open Studio[2].

Table 1 shows the results of the conceptual data model reverse engineering process related to the three case studies. It reports the number of classes and relationships that were automatically inferred from each information system.

---

[1] http://msdn.microsoft.com/en-us/library/wss56bz7.aspx.

[2] https://www.talend.com/.

Table 1. Reverse engineering results.

Information system	#Classes	#Association relationships	#Composition relationships
SW TC	49	33	48
Test pattern	36	23	35
KPI	25	12	24

After the analysis, we performed a validation step in order to assess the acceptability and the precision of the inferred models. To this aim we enrolled three experts from the company belonging to the application domains of the three information systems.

We submitted them: (1) the inferred data models, (2) a report containing the description of each inferred UML item and one or more traceability links towards the spreadsheet's parts from which it was extracted, (3) a set of questions in order to collect the expert judgments about the validity of the UML item proposals. We asked the experts to answer the questions and thus we were able to assess the effectiveness of the overall reverse engineering process by means of the Precision metric reported below:

$$Precision = \frac{|V.I.E.|}{|I.E.|} \times 100$$

I.E. is the set of the UML elements, i.e., classes, class names, class attributes, relationships between classes and multiplicities of the relationships that were inferred by the process. $|I.E.|$ is the cardinality of the I.E. set.

$V.I.E. \subseteq I.E.$ is the number of the inferred UML elements that were validated by the industrial expert and $|V.I.E.|$ is the cardinality of this set.

Table 2 shows the precision values we obtained for each information system. The precision values reported showed that more than 80 % of the inferred elements were validated by the experts. In the remaining cases, the experts proposed some minor changes to the candidate classes. As an example, with respect to the first system, the Test Patterns one, the expert decided to candidate additional classes by extracting their attributes from the ones of another candidate class, or to merge some of the proposed classes into a single one.

Table 2. Evaluation results.

Information system	Precision
SW TC	81 %
Test pattern	84 %
KPI	92 %

5 Lessons Learned

We analyzed in depth the results of the validation steps in order to assess the applicability of the rules used throughout the process. In this way, we were able to learn some lessons about the effectiveness of the heuristic rules.

In particular, with respect to the class diagram reported in Fig. 12, we observed that the expert proposed (1) to discard the *PlotOutput* class and to move its attributes into the *ExpectedResult* one and (2) to extract a new class, named *Repetition*, having the attributes *repetition* and *eventTime* that were given from the candidate class named *Test Step*. Similar changes were proposed also in the other two case studies we performed.

We observed that: (1) when the expert decided to *merge* two candidate classes into a single one, these classes had been considered as different by *Rule 4.2*, since they derived from two subareas having headers with different colors but actually belonging to the same concept. Whereas, (2) when the expert decided to *extract* an additional class from a candidate one and assigned it a subset of the attributes of the inferred one the *Rule 4.2*, *Rule 5* and *Rule 6* were not able to identify this extra class since the cells associated to these two concepts had the same formatting and layout style. As a consequence, our process was not able to discriminate between them. In both cases the process failed because the spreadsheet did not comply with the formatting rules exploited by our process.

We studied in detail the cases in which the experts proposed the changes in order to understand if it was possible to introduce new heuristics.

In the case (1) we were not able to find any common property between cells belonging to the classes that were proposed to be merged by the expert, as a consequence no new heuristic was introduced. In case (2) we observed that the expert proposed to extract the columns/rows of a candidate class into a new one when they presented a high level of data replication percentage (> 80 %). On the basis of this observation we proposed the following heuristic rule that can be applied as a further Step of the proposed process.

Step 8. In this step the *Rule 8* is exploited. This heuristic is applied to the overall corpus of spreadsheets. It analyzes the data contained in the columns/rows belonging to the classes that were inferred at the end of Step 7. If two or more columns/rows related to a given class C_i have a data replication percentage that is higher than 80 % then a new class is extracted from the original one. The attributes of the extracted class have the same name of the considered columns. The extracted class is named as follows: $Name\_of\_C_i\_Extracted_i$. A relationship association is inferred between the classes. The multiplicity of the association on the side related to the class *Ci* is equal to 1 whereas the one on the other side is 1..*. The data quality tool Talend was employed to implement this heuristic.

As an example, by applying the *Step 8* on the *Test Step* class we were able to obtain a result similar to the one proposed by the expert. In detail, the *Rule* automatically extracted a class, named *Test Step_Extracted1*, having the attributes related to the *repetition* and *eventTime* columns. The new class was inferred since the level of data replication of the two considered columns was higher than 80 %. Figure 13 shows one of the data analysis views provided by Talend. The histogram reports the results of the analysis about the data cells belonging to the columns *repetition* and *eventTime*. By applying the formula reported below we observe that in this case the percentage of data replication is equal to 93,8 %.

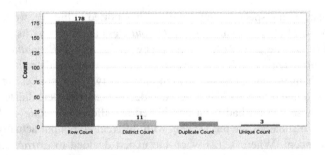

Fig. 13. Example of a histogram reporting a data replication analysis on the class Test Step of an example spreadsheet.

$$Data\ Replication\ Percentage = \frac{Row\ Count - Distinct\ Count}{Row\ Count} \times 100$$

Figure 14 shows an example of the application of the *Step 8* on the candidate class *Test Step*.

To confirm the validity of the *Rule 8,* we applied the *Step 8* to the models that were inferred by the previous process and measured the precision of the resulting class diagrams. The results reported in Table 3 show that by applying this step the effectiveness of the whole process increases.

As to the rules we used to associate the candidate classes with a name, only in 21 over 110 cases they failed, since the spreadsheets did not include meaningful information to be exploited for this aim.

Furthermore we observed that in some cases the expert decided to discard some of the proposed composition relationships between the inferred classes. This fact occurred when the proposed conceptual class diagram presented a particular pattern like the one showed in Fig. 15-A. In this case, the expert decided to move the attributes of the leaf class (*Sheet_SubArea1*) to the *Sheet1* class and to remove the remaining classes, as

Fig. 14. Example of Step 8 execution.

Table 3. New process results.

Information system	Precision
SW TC	90 %
Test pattern	92 %
KPI	95 %

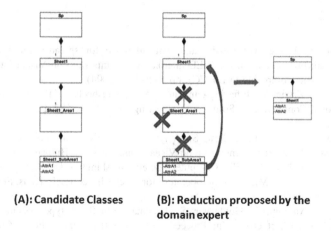

(A): Candidate Classes   **(B): Reduction proposed by the domain expert**

Fig. 15. Example of Class Diagram reduction proposed by the domain expert.

shown in Fig. 15-B. This specific pattern occurred 8 times and in 6 cases the expert decided to make these changes. This result showed us the need to introduce new rules aimed at reducing the complexity of the class diagram that may be used in the occurrence of this particular pattern.

6 Conclusions and Future Works

In this paper we presented a process for inferring a conceptual data model from a spreadsheet-based information system. The process has been defined in an industrial context and validated by an experiment involving three different spreadsheet-based information systems from the considered automotive industrial domain. The results of the experiment showed the applicability of the process and the acceptability of the inferred models, according to the judgment of experts of application domains where the spreadsheets were used. The lessons we learned from this experience allowed us to improve the early version of the process.

Our work differs from other ones described in the literature, since the proposed approach has been tailored to spreadsheets used to implement information systems, rather than calculation sheets.

In future work, we plan to perform further experimentations involving other spreadsheet-based systems belonging to other application domains. Moreover, we want to extend our approach further, by proposing reverse engineering techniques aimed at inferring the functional dependencies between the data embedded in the spreadsheets by analyzing the VBA functionalities they include.

Acknowledgements. This work was carried out in the contexts of the research projects IES-WECAN (Informatics for Embedded SoftWare Engineering of Construction and Agricultural machiNes - PON01-01516) and APPS4SAFETY (Active Preventive Passive Solutions for Safety - PON03PE_00159_3), both partially founded by the Italian Ministry for University and Research (MIUR).

References

1. Abraham, R., Erwig, M.: Header and unit inference for spreadsheets through spatial analyses. In: Proceedings of the IEEE International Symposium on Visual Languages and Human-Centric Computing (VL/HCC), pp. 165–172 (2004)
2. Abraham, R., Erwig, M.: Inferring templates from spreadsheets. In: Proceedings of the 28th International Conference on Software Engineering (ICSE), pp. 182–191. ACM, New York (2006)
3. Abraham, R., Erwig, M., Andrew, S.: A type system based on end-user vocabulary. In: Proceedings of the IEEE Symposium on Visual Languages and Human-Centric Computing (VL/HCC), pp. 215–222. IEEE Computer Society, Washington, DC (2007)
4. Abraham, R., Erwig, M.: Mutation operators for spreadsheets. IEEE Trans. Softw. Eng. **35** (1), 94–108 (2009)
5. Ahmad, Y., Antoniu, T., Goldwater, S., Krishnamurthi S.: A type system for statically detecting spreadsheet errors. In: Proceedings of the IEEE International Conference on Automated Software Engineering, pp. 174–183. (2003)
6. Amalfitano, D., Fasolino, A.R., Maggio, V., Tramontana, P., Di Mare, G., Ferrara, F., Scala, S.: Migrating legacy spreadsheets-based systems to Web MVC architecture: an industrial case study. In: Proceedings of CSMR-WCRE, pp. 387–390 (2014)
7. Amalfitano, D., Fasolino, A.R., Maggio, V., Tramontana, P., De Simone, V.: Reverse engineering of data models from legacy spreadsheets-based systems: an Industrial Case Study. In: Proceedings of the 22nd Italian Symposium on Advanced Database System, pp. 123–130 (2014)
8. Amalfitano, D., Fasolino, A.R., Tramontana, P., De Simone, V., Di Mare, G., Scala, S.: Information extraction from legacy spreadsheet-based information system - an experience in the automotive context. In: DATA 2014, pp. 389–398 (2014)
9. Bovenzi, D., Canfora, G., Fasolino, A.R.: Enabling legacy system accessibility by Web heterogeneous clients. In: Proceedings of the Seventh European Conference on Software Maintenance and Reengineering, pp. 73–81. IEEE CS Press (2003)
10. Canfora, G., Fasolino, A.R., Frattolillo, G., Tramontana, P.: A wrapping approach for migrating legacy system interactive functionalities to service oriented architectures. Elsevier, J. Syst. Softw. **81**(4), 463–480 (2008)
11. Chen, Z., Cafarella, M.: Automatic web spreadsheet data extraction. In: Proceedings of the 3rd International Workshop on Semantic Search Over the Web (SS@ 2013), p. 8. ACM, New York (2013)
12. Cunha, J., Saraiva J., Visser, J.: From spreadsheets to relational databases and back. In: Proceedings of the 2009 ACM SIGPLAN Workshop on Partial Evaluation and Program Manipulation, PEPM 2009, pp 179–188. ACM, New York (2009)
13. Cunha, J., Erwig, M., Saraiva, J.: Automatically inferring ClassSheet models from spreadsheets. In: Proceedings of the 2010 IEEE Symposium on Visual Languages and Human-Centric Computing, VLHCC 2010, pp 93–100. IEEE Computer Society (2010)
14. Cunha, J., Mendes J., Fernandes J.P., Saraiva J.: Embedding and evolution of spreadsheet models in spreadsheet systems. In: VL/HCC 2011: IEEE Symposium on Visual Languages and Human-Centric Computing, pp 186–201. IEEE Computer Society (2011)
15. Cunha, J., Fernandes, J.P., Mendes, J., Pacheco, H., Saraiva, J.: Bidirectional transformation of model-driven spreadsheets. In: Hu, Z., de Lara, J. (eds.) ICMT 2012. LNCS, vol. 7307, pp. 105–120. Springer, Heidelberg (2012)

16. Cunha, J., Fernandes, J.P., Mendes, J., Saraiva, J.: MDSheet: A framework for model-driven spreadsheet engineering. In: Proceedings of the 34rd International Conference on Software Engineering, ICSE 2012, pp 1412–1415. ACM (2012)
17. Cunha, J., Fernandes, J.P., Mendes, J., Saraiva, J.: Towards an evaluation of bidirectional model-driven spreadsheets. In: User evaluation for Software Engineering Researchers, USER 2012, pp 25–28. ACM Digital Library (2012)
18. Cunha, J., Fernandes, J.P., Saraiva, J.: From relational ClassSheets to UML+OCL. In: The Software Engineering Track at the 27th Annual ACM Symposium on Applied Computing (SAC 2012), Riva del Garda (Trento), Italy, pp. 1151–1158. ACM (2012)
19. Cunha, J., Mendes, J., Saraiva, J., Visser, J.: Model-based programming environments for spreadsheets. Sci. Comput. Program. (SCP) 96(2), 254–275 (2014)
20. Cunha, J., Fernandes, J., Mendes, J., Saraiva, J.: Embedding, evolution, and validation of model-driven spreadsheets. IEEE Trans. Softw. Eng. 41(3), 241–263 (2014)
21. Cunha, J., Erwig, M., Mendes, J., Saraiva, J.: Model inference for spreadsheets. Autom. Softw. Eng., 1–32 (2014). Springer, USA
22. De Lucia, A., Francese, R., Scanniello, G., Tortora, G.: Developing legacy system migration methods and tools for technology transfer. Softw. Pract. Experience 38(13), 1333–1364 (2008). Wiley
23. Di Lucca, G.A., Fasolino, A.R., De Carlini, U.: Recovering class diagrams from data-intensive legacy systems. In: Proceedings of International Conference on Software Maintenance, ICSM, pp. 52–62. IEEE CS Press (2000)
24. Fisher, M., Rothermel, G.: The EUSES spreadsheet corpus: a shared resource for supporting experimentation with spreadsheet dependability mechanisms. In: 1st Workshop on End-User Software Engineering, pp. 47–51 (2005)
25. Hermans, F., Pinzger, M., van Deursen, A.: Automatically extracting class diagrams from spreadsheets. In: D'Hondt, T. (ed.) ECOOP 2010. LNCS, vol. 6183, pp. 52–75. Springer, Heidelberg (2010)
26. Hermans F., Pinzger, M., van Deursen, A.: Supporting professional spreadsheet users by generating leveled dataflow diagrams. In: Proceedings of the 33rd International Conference on Software Engineering (ICSE 2011), pp. 451–460. ACM, New York (2011)
27. Hung, V., Benatallah, B., Saint-Paul R.: Spreadsheet-based complex data transformation. In: Proceedings of the 20th ACM International Conference on Information and Knowledge management (CIKM 2011), pp. 1749–1754. ACM, New York (2011)
28. Janvrin, D., Morrison, J.: Using a structured design approach to reduce risks in end user spreadsheet development. Inf. Manag. 37(1), 1–12 (2000)
29. Mittermeir, R., Clermont, M.: Finding high-level structures in spreadsheet programs. In: Proceedings of the Ninth Working Conference on Reverse Engineering (WCRE), pp. 221–232. IEEE Computer Society (2002)
30. Panko, R.R., Halverson, R.P.: Individual and group spreadsheet design: patterns of errors. In: Proceedings of the Hawaii International Conference on System Sciences (HICSS), pp. 4–10 (1994)
31. Ronen, B., Palley, M.A., Lucas, H.C.: Spreadsheet analysis and design. Commun. ACM 32, 84–93 (1989)
32. Scaffidi, C., Shaw, M., Myers, B.: Estimating the numbers of end users and end user programmers. In: 2005 IEEE Symposium on Visual Languages and Human-Centric Computing, 20–24 September 2015, pp. 207–214 (2005)
33. Shokry, H., Hinchey, M.: Model-based verification of embedded software. IEEE Comput. 42(4), 53–59 (2009)

Validation Approaches for a Biological Model Generation Describing Visitor Behaviours in a Cultural Heritage Scenario

Salvatore Cuomo[1]([⊠]), Pasquale De Michele[2], Giovanni Ponti[2], and Maria Rosaria Posteraro[1]

[1] Department of Mathematics and Applications, University of Naples "Federico II", Naples, Italy
salvatore.cuomo@unina.it
[2] ENEA – Portici Research Center, Technical Unit for ICT – High Performance Computing Division, Naples, Italy
{pasquale.demichele,giovanni.ponti}@enea.it

Abstract. In this paper we propose a biologically inspired mathematical model to simulate the personalized interactions of users with cultural heritage objects. The main idea is to measure the interests of a spectator w.r.t. an artwork by means of a model able to describe the behaviour dynamics. In this approach, the user is assimilated to a computational neuron, and its interests are deduced by counting potential spike trains, generated by external currents. The key idea of this paper consists in comparing a strengthened validation approach for neural networks based on *classification* with our novel proposal based on *clustering*; indeed, clustering allows to discover natural groups in the data, which are used to verify the neuronal response and to tune the computational model.

Preliminary experimental results, based on a phantom database and obtained from a real world scenario, are shown. They underline the accuracy improvements achieved by the clustering-based approach in supporting the tuning of the model parameters.

Keywords: Computational neural models · Clustering · Data mining · User profiling

1 Introduction

In the heritage area, the needs of innovative tools and methodologies to enhance the quality of services and to develop smart applications is an increasing requirement. Cultural heritage systems contain a huge amount of interrelated data that are more complex to classify and analyze.

For example, in an art exhibition, it is of great interest to characterize, study, and measure the level of knowledge of a visitor w.r.t. an artwork, and also the dynamics of social interaction on a relationship network. The study of individual

© Springer International Publishing Switzerland 2015
M. Helfert et al. (Eds.): DATA 2014, CCIS 178, pp. 154–168, 2015.
DOI: 10.1007/978-3-319-25936-9_10

interactions with the *tangible culture* (e.g., monuments, works of art, and arti-facts) or with the *intangible culture* (e.g., traditions, language, and knowledge) is a very interesting research field.

To understand and to analyze how artworks influence the social behaviours are very hard challenges. Semantic web approaches have been increasingly used to organize different art collections not only to infer information about an opera, but also to browse, visualize, and recommend objects across heterogeneous collections [18]. Other methods are based on statistical analysis of user datasets in order to identify common paths (i.e., *patterns*) in the available information. Here, the main difficulty is the management and the retrieval of large databases as well as issues of privacy and professional ethics [17]. Finally, models of artificial neural networks, typical of Artificial Intelligence field, are adopted. Unfortunately, these approaches seems to be, in general, too restrictive in describing complex dynamics of social behaviours and interactions in the cultural heritage framework [16].

In this paper, we propose a comparative analysis for classification and clustering approaches, in order to discover a reliable strategy to tune the model parameters. Specifically, we adopt two different strategies to discover data groups: the first one consists in exploiting the supervised data groupings by means of a bayesian classifier [8], whereas the second one is based on a new approach that finds data groupings in an unsupervised way [9]. Such a strategy resorts to a *clustering* task employing the well-known K-means algorithm [13]. The main purpose of our research has the aim of underlining the advantages of the clustering-based approach w.r.t. the one based on the bayesian classifier in producing data groups (i.e., *clusters*) that highlight hidden patterns and previously unknown features in the data. This fact naturally impacts on the following step, which consists in estimating the values characterizing neuron electrical properties of the adopted network model. Such a discovered mathematical model is particularly suitable to analyze visitor behaviours in cultural assets [3,6,7].

In this phase of our approach, we refer to a computational neuroscience terminology for which a cultural asset visitor is *a neuron* and its interests are *the electrical activity* which has been stimulated by appropriate currents. More specifically, the dynamics of the information flows, which are the social knowledge, are characterized by neural interactions in biological inspired neural networks. Reasoning by similarity, the users are the neurons in a network and its interests are the morphology; the common topics among users are the neuronal synapses; the social knowledge is the electrical activity in terms of quantitative and qualitative neuronal responses (spikes). This lead to produce a characterization of user behaviours in exhibits, starting from a real world scenario.

The paper is organized as follows. In Sect. 2 we report the preliminaries and the mathematical background of the problem. In Sect. 3 we discuss the proposed validation approaches, which based on the bayesian classifier and on clustering, respectively. Section 4 provides a comparative description of the results achieved by the two approaches, and propose the neuron model adopted, also furnishing some technical and implementation details. Section 5 is devoted to the related works. Finally, conclusions and some scratch for future works are drawn in the Sect. 6.

2   Preliminaries and Mathematical Background

Our research starts from the data collected in a real scenario. The key point event was an art exhibition within *Maschio Angioino Castle*, in Naples (Italy), of sculptures by Francesco Jerace, promoted by DATABENC [10], a High Technology District for Cultural Heritage management recently founded by Regione Campania (Italy). The sculptures was located in three rooms and each of them was equipped with a sensor, able to "talk" with the users. After the event, the collected data have been organized in a structured knowledge entity, named "booklet" [5]. The booklet contents are necessary to feed the artworks fruition and they require a particular structure to ensure that the artworks start to talk and interact with the people. The Listing 1.1 shows a XML schema diagram of a simplified model of the booklet entity, characterized by the attributes of an artwork.

We analyze the *log file* of a phantom database that was populated with both real and random data. It represents the basic knowledge on which we test the applicability of the proposed biological inspired mathematical model.

In general, a mathematical model, corresponding to a particular physical system S, consists of one or more equations, whose individual solutions, in response to a given input, represent a good approximation of the variables that are measured in S. A biological neuron model consists of a mathematical description of nervous cell properties, more or less accurate, and allows to describe and predict certain biological behaviours. A neuron can be modeled at different levels of complexity: if we consider the propagation effects, then we have compartmental models defined by means of Partial Differential Equations (PDEs); if, instead, we assume that the action potential propagation is almost instantaneous if compared to the time scale of the generation of itself, then we have single compartment models defined by means of Ordinary Differential Equations (ODEs) and algebraic equations.

In our experiments, we adopted the *Integrate & Fire (I&F)* model, which is a simple ODE scheme that considers the neuron as an electrical circuit in which only the effects of the membrane capacitance are evaluated. The circuit is represented by the time derivative of the capacitance law $(Q = CV)$, that is

$$\begin{cases} \frac{dV}{dt} + \frac{V}{\tau} = \frac{I(t)}{C} \\ V(0) = V_0 \\ \text{if} \quad \exists t : V(t) = \theta \rightarrow V(t)^+ = 0 \end{cases} \tag{1}$$

where $t^+ = t + \epsilon$ with ϵ very small, V_m is the membrane potential, C_m is the membrane capacitance, $I(t)$ is the ionic current of the neuron m, τ is the time constant

$$\tau = R \cdot C \tag{2}$$

and R_m is the resistance.

Listing 1.1. An example of the structured LOG file.

```
1  <?xml version=" 1.0"  encoding="UTF-8"?>
2   <USER ID='UI001'>
3       <STEREOTYPE_USER>2</STEREOTYPE_USER>
4       <START_SESSION></START_SESSION>
5       <END_SESSION></END_SESSION>
6    <TRANSACTION>
7     <REQUEST>
8       <HTTP_METHOD>GET</HTTP_METHOD>
9       <PATH_INFO>/opera</PATH_INFO>
10      <REQUEST_PARAMETERS>
11        <CODEARTWORK>ART0224VICTA</CODEARTWORK>
12        <DATE>13/05/2013</DATE>
13        </REQUEST_PARAMETERS>
14      <REMOTE_ADDRESS>192.168.1.6</REMOTE_ADDRESS>
15     </REQUEST>
16      <PARAMETERS_LOG>
17        <HOUR_LISTEN_START>13/05/2013  13:58:12</HOUR_LISTEN_START>
18        <HOUR_LISTEN_END>13/05/2013  14:05:42</HOUR_LISTEN_END>
19        <AUDIOS>
20      <TOT_NUMBER>3</TOT_NUMBER>
21      <AUDIO ID='AU1111'>
22          <HOUR_END>13/05/2013  14:00:42</HOUR_END>
23        <LENGTH>180</LENGTH>
24      </AUDIO>
25        </AUDIOS>
26        <IMAGES>
27      <TOT_NUMBER>11</TOT_NUMBER>
28      <IMAGE  ID='IM1122' />
29      <IMAGE  ID='IM1134' />
30      <IMAGE  ID='IM1135' />
31        </IMAGES>
32        <VIDEOS>
33      <TOT_NUMBER>2</TOT_NUMBER>
34      <VIDEO ID='VI3333'>
35          <HOUR_END>13/05/2013  14:20:12</HOUR_END>
36        <LENGTH>180</LENGTH>
37     </VIDEO>
38        </VIDEOS>
39        <TEXTS>
40      <TOT_NUMBER>4</TOT_NUMBER>
41      <TEXT ID='TX4455' />
42          <TEXT ID='TX4456' />
43          <TEXT ID='TX4457' />
44          <TEXT ID='TX4458' />
45   </TEXTS>
46        </PARAMETERS_LOG>
47    </TRANSACTION>
48  </USER>
```

The application of an external current in input leads a membrane potential increase, until this reaches a threshold value: at this point the neuron emits a spike, after which the potential V_m returns at the rest value. The *I&F* describes simplified biological dynamics able to illustrate only some features of the neuronal activities. Our goal is to apply the discussed model to a case study of an artwork visitor of a cultural heritage asset in an exhibit.

3   Validation Approaches

In this section, we show the details of the investigated validation approaches to discover neuronal network model. The first one is supervised and consists in the adoption of a naive bayesian classifier, and the second one consists in an unsupervised strategy resorting to a clustering algorithm.

3.1   Naive Bayesian Classifier

We have organized the log file structure (Listing 1.1, discussed in the Sect. 2, in a Weka's ARFF file format [21] and we have used it as an input for the bayesian classifier. In the following, we report an example of the typical adopted ARFF file.

```
@RELATION ARTWORK
@ATTRIBUTE audios {p02, p04, p06, p08, p1}
@ATTRIBUTE images {p02, p04, p06, p08, p1}
@ATTRIBUTE texts {p02, p04, p06, p08, p1}
@ATTRIBUTE spike {yes, no}
@DATA
p02,p02,p02,No
p04,p02,p02,No
...
p1,p04,p02,Yes
...
```

In the proposed scheme, the values
$$p02, \ p04, \ \ldots, \ p1$$

are the percentile of a fixed ATTRIBUTE (audios, images or texts). For example, the value $p02$ of the feature *images* means that the user has viewed almost the 20 % of overall images available for the specific artwork. A special role is played by the ATTRIBUTE spike that reports the interest about an opera w.r.t a suitable feature combination. More in details, the tuple $p1, p04, p02$ means that the user has listen between the 80 % and 100 % of the available audios, see between the 20 % and 40 % of the available images, read between the 0 % and 20 % of the available texts and, in this case, *spike* is equal to yes, i.e. user is interested to the artwork.

We recall that we are interested to find the *I&F* dynamic correlation with the output of a such well-known classification method. Then, in order to have a comparison metric with the results returned from the model, we choose to analyze the data trough a naive bayesian classifier. The selected one is fairly intuitive and is based on the minimization of the following cost function:

$$CM(x_1, x_2, \cdots, x_n) = \arg\max_z p(Z = z) \prod_{i=1}^{N} p(X_i = x_i | Z = z)$$

where Z is a dependent class variable and $X_1 \cdots X_n$ are several feature variables. The classifier is based on the computation of individual conditional probabilities for each values of the class variable Z and for each feature $p(X_i|Z_j)$. The class, given by bayesian classifier, is the one for which we have the largest product of the probabilities. The Maximum Likelihood Estimation Method [20] is used to determine the individual conditional probabilities.

In a first step we use the bayesian classifier for investigate the data and in the Table 1 we report some output. In this way, combining the values of the attributes audios, images and texts, it is possible to obtain a total of $N = 125$ different tuples, belonging to the set called U. Assuming split these tuples into two classes: let be C the class of the tuples that involve a "spike", namely the class of tuples to which the value yes, of the attribute spike, is associated; let be $U \setminus C$ the class of the tuples that do not involve a "spike", namely the class of tuples to which the value no, of the attribute spike, is associated.

Table 1. BC classifier metrics.

# of elements to classify	125
# of True positive	65
# of False positive	0
# of True negative	52
# of False negative	8
Precision	1
Recall	0.89

At the end of the classification process, on all the N elements, for which the actual classification is known, it is possible to define the following parameters:

- **True Positive** (TP), that is the number of elements that were classified as belonging to C and actually belong to C.
- **False Positive** (FP), that is the number of elements that were classified as belonging to C but that, in reality, belong to $U \setminus C$.
- **True Negative** (TN), that is the number of elements that were classified as belonging to $U \setminus C$ and actually belong to $U \setminus C$.
- **False Positive** (FP), that is the number of elements that were classified as belonging to $U \setminus C$ but that, in reality, belong to C.

Obviously, $TP + FP + TN + FN = N$ and the number of elements that the classification process has classified as belonging to C is $TP+FP$. Then, it is possible to define the two metrics needed to evaluate the quality of the classification: **recall** and **precision**, that are

$$\frac{TP}{(TP + FN)} \qquad \frac{TP}{(TP + FP)}$$

Note that both precision and recall ranges in $[0..1]$. All the above mentioned parameters are reported in the Table 1.

3.2   Clustering

In this section, we propose a new strategy to discover classes in the data which can be used for the next modeling step, that is the tuning of the electrical parameters for the circuit model characterizing the neuron. In fact, classification algorithms have the major limitation of labeling data according to a yet-known training set, as they are supervised approaches. In many real world datasets, data objects do not typically have assigned class membership, and this may lead to have accuracy issues in the whole classification process.

For this reason, we propose to address such an issue by introducing a *clustering*-based approach [12–14] to discover data groups. Clustering is an *unsupervised* task, since it can be applied to unclassified data (i.e., unlabeled) to obtain homogeneous object groupings. In this approach, groups are more representative w.r.t. single object as they summarize their common features and/or patterns; indeed, objects belonging to the same group are quite similar each other, whereas objects in different groups are quite dissimilar.

In our context, data to be clustered are *tuples* representing visitor's behaviours related to an artwork. Note that now "spike" has a more informative role in the dataset, as it is not seen as a class but as a further information about visitor's behaviour. In our experiments, we assume the following criteria for spike generation. A visitor enjoyed an artwork if he benefits from the whole content of at least one of the available services, or if he exploits more than the 66 % of the total contents.

This new clustering-based approach allows us to produce a more general dataset, in which we do not need to assign object classes, and also attributes can take values in a continuous range, instead of in a discrete one. Therefore, the clustering phase produces groups according to visitor's preferences, which are not necessary driven by spike generation.

On these hypothesis, we have rearranged the log file structure (Listing 1.1), in a suitable way for clustering process, as follow:

```
@RELATION ARTWORK
@ATTRIBUTE audios NUMERIC [0..1]
@ATTRIBUTE images NUMERIC [0..1]
@ATTRIBUTE texts NUMERIC [0..1]
@ATTRIBUTE spike {0,1}
@DATA
0.1,0.4,1.0,1
0.3,0.6,0.4,0
. . .
0.5,1.0,0.7,1
. . .
```

In the proposed scheme, data values represent the amount of information that the visitor has exploited for an artwork for each attribute of the dataset, and the last attribute describes the spike generation according to the algorithm previously described. In this way, combining the values of the attributes `audios`, `images` and `texts`, it is possible to obtain a total of $N = 1,331$ different data objects (i.e., tuples) — for simplicity, we take into account just real values rounded at the first decimal value.

As regards the clustering task, we can employ any algorithm to discover groups. However, in this paper, we resorted to the well-known K-means clustering algorithm [13]. K-means requires only one parameter, that is the number K of clusters (i.e., groups) to be discovered. Algorithm 1 shows the outline of the K-means clustering algorithm.

In our experiments, we first started with $K = 2$, which is the natural starting choice to model a classification-like approach (i.e., "spike" or "no-spike"). Nevertheless, we can also perform further experiments by setting higher values for K to capture finest similarities and/or hidden patterns in the data. Figure 1 shows the output of the clustering phase with $K = 2$. Note that we do not take into account the "spike" attribute in the clustering process, as it could clearly bias the entire process. However, we exploited it at the end of the clustering phase to assess the result accuracy. We resorted to Weka "simpleKMeans" implementation, and the plot is also obtained employing Weka clustering visualization facilities.

Algorithm 1. K-means.

Require: a dataset objects $\mathcal{D} = \{o_1, \ldots, o_N\}$; the number of output clusters K
Ensure: a set of clusters $\mathcal{C} = \{C_1, \ldots, C_K\}$
 1: **for** $i = 1$ to K **do**
 2:    $c_i \leftarrow$ randomInitialize($\mathcal{D}$)
 3: **end for**
 4: **repeat**
 5:    **for all** $C_i \in \mathcal{C}$ **do**
 6:      $C_i \leftarrow \emptyset$
 7:    **end for**
 8:    **for all** $o_u \in \mathcal{D}$ **do**
 9:      $j \leftarrow argmin_{i \in [1..K]} dist(o_u, c_i)$
10:      $C_j \leftarrow C_j \cup \{o_u\}$
11:    **end for**
12:    **for all** $C_i \in \mathcal{C}$ **do**
13:      $c_i \leftarrow$ updateCentroid(C_i)
14:    **end for**
15: **until** centroids do not change or a certain termination criterion is reached

---

The plot represents tuples in terms of cluster membership (x-axis) and spike emission (y-axis). It is easy to note that all the data in *cluster0* refer to tuples that produce spikes (i.e., with value 1), whereas all the ones in *cluster1* identify tuples that do not emit spike (i.e., with value 0). Therefore, evaluating clustering results in terms of well-separation of the data w.r.t. the spike emission issue, we achieved a high-quality clustering as all the data have been correctly separated.

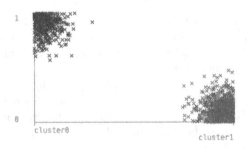

Fig. 1. Clustering results for K-means ($K = 2$).

4 Comparative Analysis and the Neuronal Model

By comparing the two proposed approaches, it is easy to note that the one based on clustering furnishes more significant groups, which are more homogeneous in terms of spike emission. In fact, summary results for the bayesian-based model are shown in Table 1, which highlights a certain number of False Negative data (i.e., $FN = 8$). Consequently, this fact negatively affects Recall value, that is equal to 0.89. On the contrary, the clustering-based approach does not assign any false membership in the data. This naturally impacts on our purpose of identifying user behaviours. For these motivations, we adopted the clustering task to guide model parameter identification.

Then, starting from the clustering output, we have integrated the *I&F* computational model in order to find some correlations with the clustering results. In particular, the couple (R, C) represents the visitor sensitivity to the artwork. We have exploited the clustering results in order to tune the values of the resistance R and conductance C of the circuit that represents the model. In a first experiment, a good choice for the couple (R, C) is

$$(R, C) = (0.51\,\text{kOhm}, 30\,\mu\text{F})$$

The current is a linear combination of the values of the attributes in the dataset. The Fig. 2 gives the dynamic response of the neuron.

In the first case (top of the Fig. 2) the current $I(t)$ is not sufficient to trigger a potential difference which gives a spike. In the second one (bottom of the Fig. 2) the neuron that has received stimuli is able to produce an interesting dynamic.

In these experiments, we show how the computational model and the clustering give information about the interest of a visitor about an artwork. In the Table 2, experimental results for the clustering and our model are reported. M.C.F. represents the Media Content Fruition w.r.t. the overall media contents. With the symbol (*) we have labeled the tuple combinations that contain the information about the fully fruition of at least one media content. Note that the last column of the table indicates the degree of the visitor interest for an artwork. Thus, in this respect, such an information is obtained by the proposed *I&F* neuron model to achieve a fine-grained indication for spikes.

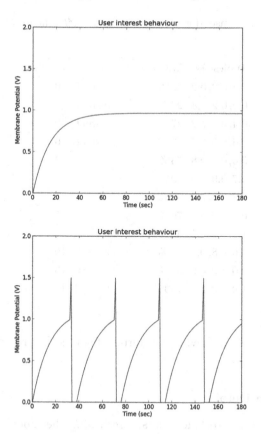

Fig. 2. Top. With a current $I(t) = 0.6 + 0.6 + 0.7$, we observe no spike presence.
Bottom With a current $I(t) = 0.6 + 0.8 + 0.8$ we observe 4 spikes.

In the Fig. 3, we have fixed

$$I(t) = 0.8 + 0.9 + 0.3$$

as a stimulus and we have compared two users U_1 with $(R, C) = (0.51, 30)$ and U_2 with $(R, C) = (0.6, 28)$.

We can observe the different number of spikes between U_1 and U_2 respect to the same artwork. If the spike are related to the the interests that a cultural asset has aroused in a viewer, the $I \& F$ is able to emerge this features. The choice of the pair (R, C) suitable for a established user is the real challenge of the model. More in general, it may be multiple scenarios to apply these dynamics. An example is the case of a cultural asset exhibition in which the target is how to place artworks. A possible choice is to select the operas that have attracted the visitors with common interests, i.e., users with similar (R, C). In the context-aware profiling instead the aim is how to change (R, C) in such a

Table 2. Spike response for clustering and *I&F* neuron with $(R, C) =$ $(0.51\,\text{kOhm}, 30\,\mu\text{F})$.

Tuples	M.C.F. (%)	Cluster	# spikes
$0.2, 0.2, 0.2$	20 %	cluster1	0
$0.2, 0.2, 0.4$	27 %	cluster1	0
$0.4, 0.2, 0.2$	27 %	cluster1	0
$0.6, 0.6, 0.7$	63 %	cluster1	0
$0.6, 0.8, 0.8$	73 %	cluster0	4
$0.7, 0.9, 0.5$	70 %	cluster0	4
$0.8, 0.9, 0.3$	67 %	cluster0	2
$0.8, 0.9, 0.6$	76 %	cluster0	5
$1.0, 0.2, 0.1$	43 %$^{(*)}$	cluster0	5
$1.0, 0.8, 0.9$	90 %$^{(*)}$	cluster0	10
$1.0, 1.0, 0.6$	86 %$^{(*)}$	cluster0	13
$1.0, 1.0, 1.0$	100 %$^{(*)}$	cluster0	16

way to predict the user behaviours in terms of spikes that represent its cultural assets.

Implementation Details

Here, we provide some technical details regarding the model implementation. From Sect. 2, rearranging the Eq. (1) and replacing the Eq. (2) in this, we obtain

$$\frac{dV}{dt} = -\frac{V}{RC} + \frac{I(t)}{C} \tag{3}$$

Then, replacing $dV = V' - V$ into the Eq. (3) we have

$$V' = V + \left(-\frac{V}{RC} + \frac{I(t)}{C} \right) dt \tag{4}$$

Finally, from the the Eq. (4), we obtain

$$V' = V + \left(\frac{-V + I(t)R}{RC} \right) dt \tag{5}$$

The Eq. (5) is the core of our model, as we can observe in Fig. 4, where V[i] is V', V[i-1] is V and I_t is $I(t)$.

The model is implemented in Python (v. 2.7.8), and we resorted to a set of Python libraries, that are math for the mathematical functions, numpy for the list management (array), and pylab for the plotting.

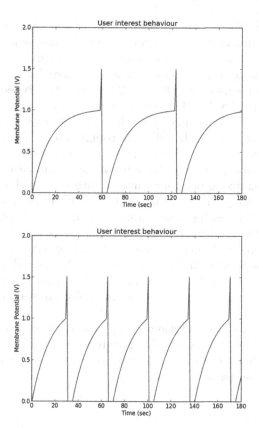

Fig. 3. Top. With the couple $(R, C) = (0.51, 30)$ the neuron has 2 spikes. **Bottom** With the couple $(R, C) = (0.6, 28)$ the neuron has 5 spikes.

```
if threshold > threshold_rest:
    V[i] = V[i-1] + (-V[i-1] + I_t*R) / tau * dt
if V[i] >= V_threshold:
    V[i] += V_spike
    threshold_rest = threshold + tau_ref
```

Fig. 4. Python code.

5   Related Work

The studying of efficient methods for learning and classifying of the user behaviors and dynamics in the real or digital life is a very large and fascinating research area. The challenge is to define automatic frameworks based on sensor networks, semantic web models, reputation systems and classifiers able to map human activity and social user interactions. More in details a smart system should be have the ability to automatically infer interests of users and track the propagation of the information. For real life applications, in [2,4] a wireless sensor network, using Bluetooth technology, able to sense the surrounding area for detecting user devices' presence in a museum is discussed. About the digital user behaviors a study of the relevance of feedbacks, typically adopted for the profiling during long-term modeling is given in [15]. In [22] an algorithm based on the descriptors representation is developed to acquire high accuracy of recognition for long-term interests, and to adapt quickly to changing interests in the learning user activity. Other methodologies, based on the computational approaches, are based on machine-learning [11]. Here, the focus is the estimate of the dynamics of the users' group membership and the character of their social relationships that are analyzed by characterizing behavior patterns with statistical learning methods. These methods are adopted to examine the users' current behavior and to classify different kinds of relationships such as workgroup, friendship and so forth. In [19], using the users data to model an individual behavior as a stochastic process, the authors show a framework that predicts the future activity, obtained by modeling the interactions between individual processes. Ontological methodologies for user profiling in recommender systems are described in [18]. Finally, a multimedia recommender system based on the social choice problem has been recently proposed in [1].

We propose a novel computational approach based on a mathematical biological inspired model. Rather than investigate the behavior of users on the web or in digital situations, our work is focused on the track and forecast the user dynamics in a real scenario.

6   Conclusions

We described a framework that reflects the computational methodology adopted to infer information about visitors in a cultural heritage context. The challenge here is to map, in a realistic way, the biological morphology of a neuron in this application scenario. We deal with a model where the (R, C) couple represents the sensitivity of the user respect to an artwork.

We compared two different strategies for tuning model parameters, in order to find an accurate approach that is able to provide the best setting for the neuronal model. In this respect, we showed experimental results for classic bayesian classifier and new clustering methodology to obtain starting groups from which these electrical parameters can be tuned. From our experiments, it has been highlighted that clustering task is able to produce a more accurate setting.

In future research lines, we will study more complex neuronal dynamics by morphology point of view with the aim to develop models that are more close to the real users. Other research tracks will be the building of computational neural networks able to reproduce the interactions in social cultural heritage networks. In addition, regarding the preliminary clustering phase, we will tune our model with more than two clusters, with the aim of obtaining fine-grainer clustering solutions that are able to capture and to highlight other neuron aspects, apart from spike generation.

References

1. Albanese, M., d'Acierno, A., Moscato, V., Persia, F., Picariello, A.: A multimedia recommender system. ACM Trans. Internet Technol. **13**(1), 3:1–3:32 (2013)
2. Amato, F., Chianese, A., Mazzeo, A., Moscato, V., Picariello, A., Piccialli, F.: The talking museum project. Procedia Comput. Sci. **21**, 114–121 (2013)
3. Bianchi, D., De Michele, P., Marchetti, C., Tirozzi, B., Cuomo, S., Marie, H., Migliore, M.: Effects of increasing CREB-dependent transcription on the storage and recall processes in a hippocampal CA1 microcircuit. HIPPOCAMPUS **24**(2), 165–177 (2014)
4. Chianese, A., Marulli, F., Moscato, V., Piccialli, F.: SmARTweet: a location-based smart application for exhibits and museums. In: Proceedings - 2013 International Conference on Signal-Image Technology and Internet-Based Systems, SITIS 2013, pp. 408–415 (2013)
5. Chianese, A., Marulli, F., Piccialli, F., Valente, I.: A novel challenge into multimedia cultural heritage: an integrated approach to support cultural information enrichment. In: Proceedings - 2013 International Conference on Signal-Image Technology and Internet-Based Systems, SITIS 2013, pp. 217–224 (2013)
6. Cuomo, S., De Michele, P., Chinnici, M.: Parallel tools and techniques for biological cells modelling. Buletinul Institutului Politehnic DIN IASI, Automatic Control and Computer Science Section, pp. 61–75. LXI (2011)
7. Cuomo, S., De Michele, P., Piccialli, F.: A performance evaluation of a parallel biological network microcircuit in neuron. Int. J. Distrib. Parallel Syst. **4**(1), 15–31 (2013)
8. Cuomo, S., De Michele, P., Posteraro, M.: A biologically inspired model for describing the user behaviors in a cultural heritage environment. In: SEBD 2014, 22nd Italian Symposium on Advanced Database Systems, Sorrento Coast, 16th–18th June 2014
9. Cuomo, S., De Michele, P., Ponti, G., Posteraro, M.: A clustering-based approach for a finest biological model generation describing visitor behaviours in a cultural heritage scenario. In: DATA 2014 - Proceedings of 3rd International Conference on Data Management Technologies and Applications, pp. 427–433 (2014)
10. DATABENC, High Technology District for Cultural Heritage. http://www.databenc.it
11. Domingos, P.: A few useful things to know about machine learning. Commun. ACM. **55**(10), 78–87 (2012)
12. Hartigan, J.A.: Clustering Algorithms. Applied Statistics. Wiley, New York (1975)
13. Jain, A.K., Dubes, R.C.: Algorithms for Clustering Data. Prentice-Hall, Upper Saddle River (1988)
14. Kaufman, L., Rousseeuw, P.J.: Finding Groups in Data: An Introduction to Cluster Analysis. Wiley, New York (1990)
15. Kelly, D., Teevan, J.: Implicit feedback for inferring user preference: a bibliography. SIGIR Forum. **37**(2), 18–28 (2003)
16. Kleinberg, J.: The convergence of social and technological networks. Commun. ACM **51**(11), 66–72 (2008)
17. Kumar, R., Novak, J., Tomkins, A.: Structure and evolution of online social networks. In: Yu, P.S., Han, J., Faloutsos, C. (eds.) Link Mining: Models, Algorithms, and Applications, pp. 337–357. Springer, New York (2010). http://dx.doi.org/10.1007/978-1-4419-6515-8_13

18. Middleton, S.E., Shadbolt, N.R., De Roure, D.C.: Capturing interest through inference and visualization: ontological user profiling in recommender systems. In: Proceedings of the 2nd International Conference on Knowledge Capture, pp. 62–69 (2003). ISBN: 1-58113-583-1
19. Pentland, A.S.: Automatic mapping and modeling of human networks. Physica A. **378**(1), 59–67 (2007)
20. Roderick, J.A., Little, R.J.A., Rubin, D.B.: Statistical Analysis with Missing Data. Wiley Editor, New York (2002). ISBN: 978-0-471-18386-0
21. Weka, Data Mining Software in Java. http://www.cs.waikato.ac.nz/ml/weka/
22. Widyantoro, D.H., Ioerger, T.R., Yen, J.: Learning user interest dynamics with a three-descriptor representation. J. Am. Soc. Inf. Sci. Technol. **52**(3), 212–225 (2001)

A Method for Topic Detection in Great Volumes of Data

Flora Amato[1], Francesco Gargiulo[1]([envelope]), Alessandro Maisto[2],
Antonino Mazzeo[1], Serena Pelosi[2], and Carlo Sansone[1]

[1] Dipartimento di Ingegneria Elettrica e delle Teconolgie dell'Informazione (DIETI),
University of Naples Federico II, Via Claudio 21, Naples, Italy
{flora.amato,francesco.grg,mazzeo,carlosan}@unina.it
[2] Dipartimento di Scienze Politiche, Sociali e della Comunicazione (DSPSC),
University of Salerno, Snc, Via Giovanni Paolo II, Fisciano (SA), Italy
{amaisto,spelosi}@unisa.it

Abstract. Topics extraction has become increasingly important due to
its effectiveness in many tasks, including information filtering, information retrieval and organization of document collections in digital libraries.
The Topic Detection consists to find the most significant topics within a
document corpus. In this paper we explore the adoption of a methodology of feature reduction to underline the most significant topics within
a *document corpus*. We used an approach based on a clustering algorithm (X-means) over the $tf - idf$ matrix calculated starting from the
corpus, by which we describe the frequency of terms, represented by
the columns, that occur in the documents, represented by the rows. To
extract the topics, we build n binary problems, where n is the numbers of clusters produced by an unsupervised clustering approach and
we operate a *supervised* feature selection over them, considering the top
features as the topic descriptors. We will show the results obtained on
two different corpora. Both collections are expressed in Italian: the first
collection consists of documents of the University of Naples Federico II,
the second one consists in a collection of medical records.

Keywords: Topic detection · Clustering · $tf - idf$ · Feature reduction

1 Introduction

Topic extraction from texts has become increasingly important due to its effectiveness in many tasks, including information retrieval, information filtering and
organization of document collections in digital libraries.

In this paper we explore the adoption of a methodology of feature extraction and reduction to underline the most significant topic within a *corpus* of
documents.

We used an approach based on a clustering algorithm (X-means) over the
$TF - IDF$ (Term Frequency - Inverse Document Frequency) matrix calculated
starting from the corpus.

© Springer International Publishing Switzerland 2015
M. Helfert et al. (Eds.): DATA 2014, CCIS 178, pp. 169–181, 2015.
DOI: 10.1007/978-3-319-25936-9_11

In the proposed approach, each cluster represents a topic. We characterize each one of them through a set of *words* representing the documents within a cluster.

In literature a standard method for obtaining this kind of result is the usage of a non-supervised feature reduction method, such as the Principal Component Analysis (PCA). However the high dimensionality of the feature vectors associated to each document make this method not feasible.

To overcome this problem we build n binary problems, where n is the numbers of clusters produced by the X-means. Then, we operate a *supervised* feature selection over the obtained clusters, considering the cluster membership as the class and the selected top features (*words*) as the researched topic.

We exploit two different *ground truth* made by domain expert in order to evaluate the most significant group of features that separate the class of interest from all the rest.

The paper is organized as follows. Section 2 describes the most recent related work for topic detection. Section 3 outlines the proposed methodology. In Sect. 4, we present the performed experiments, showing dataset used for experimental validation and the obtained results. Finally, in Sect. 5 we discuss conclusions and possible future directions for our research.

2   Related Works

Although nowadays it is available a large amount of information on web, there is the need of accessing this information in the shortest time and with the highest accuracy. To help people to integrate and organize scattered information, several machine learning approaches have been proposed for the Topic Detection and Tracking technology (TDT). Moreover, topic detection and tracking consists in studying how to organize information, which is based on event in natural language information flow. Topic Detection (TD) is a sub-task of TDT. In the TD task the set of most prominent topics have been found in a collection of documents. Furthermore TD is based on the identification of the stories in several continuous news streams, which concern new or previously unidentified events. Sometimes unidentified events have been retrieved in an accumulated collection ("retrospective detection") while in the "on-line detection" the events could be labelled when they have been flagged as new events in real time. Give the TD is a problem to assign labels to unlabelled data grouping together a subset of news reports with similar contents, most unsupervised learning methods, proposed in literature as [1,2], exploit the text clustering algorithms to solve this problem.

Most common approaches, as [1], given list of topics, the problem of identifying and characterizing a topic is a main part of the task. For this reason a training set or other forms of external knowledge cannot be exploited and the own information contained in the collection can be used to solve the Topic Detection problem. Moreover the method, proposed in [1], is a two-step approach: in the former a list of the most informative keywords have been extracted; the latter consists in the identification of the clusters of keywords for which a center has been defined as the representation of a topic. The authors of [1] considered topic detection without any prior knowledge of the category structure or possible categories. Keywords are extracted and clustered based on different similarity measures using the induced k-bisecting clustering algorithm. They considered distance measures between words, based on their statistical distribution over a corpus of documents, in order to find a measure that yields good clustering results.

In [3] is proposed a generative model based on Latent Dirichlet Allocation (LDA) that integrates the temporal ordering of the documents into the generative process in an iterative fashion called Segmented Author-Topic Model (S-ATM). The document collection has been split into time segments where the discovered topics in each segment has been propagated to influence the topic discovery in the subsequent time segments. The document-topic and topic-word distributions learned by LDA describe the best topics for each document and the most descriptive words for each topic. An extension of LDA is the Author-Topic Model (ATM). In ATM, a document is represented as a product of the mixture of topics of its authors, where each word is generated by the activation of one of the topics of the document author, but the temporal ordering is discarded. S-ATM is based on the ATM and extends it to integrate the temporal characteristics of the document collection into the generative process. Besides S-ATM learns author-topic and topic-word distributions for scientific publications integrating the temporal order of the documents into the generative process.

The goals in [4] are: (i) the system should be able to group the incoming data into a cluster of items of similar content; (ii) it should report the contents of the cluster in summarized human-readable form; (iii) it should be able to track events of interest in order to take advantage of developments. The proposed method has been motivated by constructive and competitive learning from neural network research. In the construction phase, it tries to find the optimal number of clusters by adding a new cluster when the intrinsic difference between the presented instance and the existing clusters is detected. Then each cluster moves toward the optimal cluster center according to the learning rate by adjusting its weight vector.

In [5], a text clustering algorithm C-KMC is introduced which combined Canopy and modified k-means clustering applied to topic detection. This text clustering algorithm is based on two steps: in the former, namely C-process, has been applied Canopy clustering that splits all sample points roughly into some overlapping subsets using inaccurate similarity measure method; in the latter, called K-process, it has been employed a modified K-means that take X-means algorithm to generate rough clusters from the canopies which share common instance. In this algorithm, Canopies are an intermediate result which can reduce the computing cost of the second step and make it much easier to be used, although Canopy is not a completed cluster or topic.

The authors of [6] used Vector Space Model (VSM) to represent topics, and then they used K-means algorithm to do a topic detection experiment. They studied how the corpus size and K-means affect this kind of topic detection performance, and then they used TDT evaluation method to assess results. The experiments proved that optimal topic detection performance based on large-scale corpus enhances by 38.38 % more than topic detection based on small-scale corpus.

3   Methodology

The proposed methodology, aiming to extract and reduce the features characterizing the most significant topic within a corpus of documents, is depicted in Fig. 1.

The considered features are computed on the basis of the *Term Frequency-Inverse Document Frequency* $(tf - idf)$ matrix from the corpus.

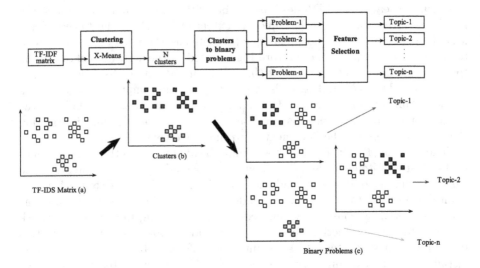

Fig. 1. Main schema.

The Process that implements the methodology is composed by several steps.

A preprocessing phase that include Text Tokenization and Normalization. In particular Text Tokenization removes the punctuation, splits the document up by blank spaces, puts every token in lowercase and removes stop words. Normalization, instead, aims at reducing the entropy of the input data, by eliminating numbers, unifying special characters, disambiguating sentence boundaries and identifying abbreviations and acronyms.

A subsequent step consists in the evaluation of a $tf-idf$ matrix [7], composed by an $m \times n$ matrix where m is the number of documents in the collection and n is the number of *tokens* we are considering. The tokens could be replaced by Lemmas, Synonymous or more complex Lexical Structures [8,9].

Each row represents a document and each column represents the $tf - idf$ value calculated for each document's token.

The $tf - idf$ value is defined as follows:

$$tf - idf = tf \times idf \qquad (1)$$

where tf is the number of the token occurrences and idf is a measure of whether the occurrence is common or rare across all collection items.

Each row of $tf - idf$ matrix contains a vector of real numbers which represents the projection of each document in an n-dimensional in feature space. For seek of simplicity, in the Fig. 1(a) we represent the documents as projected in a bi-dimensional space.

One of the main motivation to reduce the original feature space is the *curse of dimensionality* [10] and the growth of complexity in the classification model. More generally, the curse of dimensionality is the expression of all phenomena that appear with high-dimensional data, and that have often unfortunate consequences on the behaviour and performances of learning algorithms.

We adopted a novel solution based on a preliminary documents clustering in order to overcome this problem (more than 60000 tokens), that makes the usage of an unsupervised feature selection method, such as *Principal Component Analysis* (PCA), not feasible. In the Fig. 1(b) is represented the clustering process.

After that we built n binary problems where n is the number of clusters obtained. Each problem is created using a *one vs all* strategy. For example, for each cluster i all the documents that belong to it are labelled as *True*, otherwise all the other documents are labelled as *False*, see Fig. 1(c).

In order to obtain the topics characterizing each binary problem, we used a supervised feature selection method on each of them. We use the previous assigned labels indicating the group membership as class on which evaluate each feature. The feature selection gave in output a set of ranked features that represent the tokens that more discriminate the cluster under analysis from all the rest. These tokens figure out the searched topic.

In the proposed approach, the first step consists in grouping documents [11, 12]; so that, the elements appearing in the same group are more related to each other than to the ones occurring in other groups. This operation is carried out through a clustering algorithm that groups the original data into N clusters.

In this work we use X-Means, an improved version of the k-means algorithm, that allows to automatically estimate the number of clusters.

After this phase, each document is associated with a label that shows which cluster it belongs to. Therefore, if such label is taken as the class of the dataset, the approach shifts from an unsupervised to a supervised one.

Finally, we perform feature selection on each binary problem in order to extract the most significant terms that represent the related cluster. Applying the feature selection on the single problem we are able to identify the features that best discriminate the documents in the cluster. These features will describe the cluster topic.

4   Results

In order to evaluate the effectiveness of this approach, we used the Weka library [13], which is a collection of data mining algorithms.

We selected for the clustering process the X-means algorithm [14]. The X-means is an evolution of the K-means, the main difference between them is on the choice of the optimal number of clusters to use, while X-means set automatically this number, the K-means use a manual parameter (**K**).

More in details we configure the X-means with: 4 as *Max Number of Iteration* and 15 as *Max Number of Clusters*.

The proposed approach has been applied on two different corpora. Both collections are expressed in Italian: the first one, called **UNINA**, consists of documents of the University of Naples Federico II, the second one, called **Medical Records**, consists in a collection of medical records.

We compare the performances obtained on these two corpora in terms of *precision, recall* and *cluster coverage*.

$$precision = \frac{N_{out} - N_{false}}{N_{out}} \tag{2}$$

$$recall = \frac{N_{out} - N_{false}}{N_{ins}} \tag{3}$$

$$coverage = \frac{N_{out}}{N_{tot}} \tag{4}$$

where N_{out} is the number of instances in result clusters, N_{false} is the number of mistake instances in result clusters, N_{ins} is the sample size of the category under test and N_{tot} is the total number of instances. As regards the N_{false}, it consists in the number of documents that differ from the cluster modal value.

In detail, the Precision specifies the homogeneity of the cluster with reference to the documents contained in it. Moreover, the Recall refers to the percentage of documents of the same category, which is assigned to the cluster, with reference to the total number of documents of the same category (e.g. the cluster 0, that belongs to *DIDATTICA*, contains the 62.1 % of the documents of the category *DIDATTICA* that corresponds to its Recall value). Finally, the Coverage is the percentage of the total amount of documents that belong to the cluster.

For clustering we use *weka.clusterers.XMeans* with *weka.core.Euclidean Distance* as distance function.

Once identified the cluster, we use *weka.filters.unsupervised.attribute. NominalToBinary* to transform the nominal attribute containing the cluster membership into N binary attributes, i.e. one per cluster, each of which is then used as class in a binary problem.

In the end we performed the feature selection on each binary problem using *weka.attributeSelection.CorrelationAttributeEval* as attribute evaluator and *weka.attributeSelection.Ranker* as search method.

The first evaluates the worth of an attribute by measuring the Pearson's correlation between it and the class, while *Ranker* generates a ranking of the attributes by their individual evaluations.

4.1 Corpora Description

The corpus **UNINA** was originally collected from the web site of the University of Naples Federico II (Unina) and labelled by domain experts. It consists of a total of 469 documents, divided into four categories statistically characterized in the Table 1.

1. *Administrative* (Amministrativo), contains all technical/administrative documents, like scope statements;
2. *Advises* (Bandi) includes announcements, like competition notices, scholarship, and so on;
3. *Academic* (Didattica) contains academic regulations, student guides, documents for the establishment of master or courses with related rules, and departments regulations;
4. *Activities Reports* (Relazioni Attività) includes final reports of evaluation procedures, and records of meetings of the boards of the departments.

Table 1. Categories distribution for corpus **UNINA**.

Class	Number of documents
Amministrativo	17
Bandi	111
Didattica	224
Relazioni Attività	117
TOT	469

The corpus **Medical Records** consists of about 5000 medical diagnoses coming from various health care organization in Campania (Italy). Since each fragment is a document produced from Medical Center, containing a description of a patient health care. For privacy issues, these documents are opportunely anonymized. The medical records are divided into seven categories detailed within the Table 2.

1. *Consulting* (Consulenze), includes reports of generic and specialistic medical advice;
2. *Doppler* (Doppler), contains reports related to Doppler-based medical analysis, with the exception of Echocardiography;
3. *Echocardiography* (Ecoc), includes reports of Doppler-based cardiac analysis;
4. *Ultrasonography* (Ecografia), contains reports of ultrasound-based cardiac analysis;

5. *Endoscop* (Endoscopia), contains reports related to Endoscopic analysis;
6. *Operation* (Intervento), includes: preliminary diagnostic tests, procedures of preparation, anaesthetic card, composition of the team, time and date, as well as the main phases: preparation, execution, conclusion, observation and post-intervention.
7. *Radiology* (Radiologia), contains reports of radiology analysis.

Table 2. Categories distribution for corpus **Medical records.**

Class	Number of documents
Consulenze	97
Doppler	593
Ecoc	702
Ecografia	1171
Endoscopia	365
Intervento	346
Radiologia	1726
TOT	5000

4.2   Experimental Results

The results of the two case studies are reported in Tables 3 and 4.

The Table 3 defines the precision, the recall and the coverage values of each category of the **UNINA** corpus, that has been expanded to show the cluster that represented it. E.g. the category *DIDATTICA* is composed of 8 clusters, while *RELAZIONI* and *BANDI* are both made up by just one cluster. Table 4, similarly, represents significant features of the **Medical Record** corpus. As happens in the first corpus, in the second one the categories *RADIOLOGIA* and *ECOGRAFIA* have been split respectively into 6 and 2 clusters, while the categories *ECOC* and *DOPPLER* correspond exactly with one cluster.

The first consideration is that not all the classes are characterized with a cluster, this fact is directly correlated to the number of documents that belong to each of them.

For example, the absence of the class *Amministrativo* within the Corpus *UNINA* is due to the few documents that belong to that class, see Table 1. For the same motivation it is possible to justify the absence of the classes *Consulenze*, *Intervento* and *Endoscopia* in the corpus *Medical Records*, see Table 2.

Another important result is that the cluster with the highest coverage is composed in large part by noise. A possible motivation could be that the biggest cluster contains a collection of documents that are very close one another, even if they belong to different categories. A possible solution is to not consider this

Table 3. Clusters assignments for Unina: the table shows the precision, the recall and the coverage values of each category of the **UNINA** corpus.

Assigned category	Precision	Recall	Coverage
DIDATTICA	**67.0 %**	**99.5 %**	**71.1 %**
0	67.5 %	62.1 %	43.9 %
1	41.1 %	13.4 %	15.6 %
2	100.0 %	12.5 %	6 %
3	100.0 %	4.0 %	1.9 %
4	100.0 %	3.1 %	1.5 %
5	100.0 %	2.2 %	1.1 %
6	100.0 %	1.8 %	0.9 %
7	100.0 %	0.4 %	0.2 %
RELAZIONI	**99.0 %**	**81.2 %**	**20.5 %**
8	99.0 %	81.2 %	20.5 %
BANDI	**100.0 %**	**36.0 %**	**8,5 %**
9	100.0 %	36.0 %	8.5 %

Table 4. Clusters assignments for medical records: the table shows the precision, the recall and the coverage values of each category of the **Medical Records** corpus.

Assigned category	Precision	Recall	Coverage
RADIOLOGIA	**64.3 %**	**91.6 %**	**49.3 %**
0	41.0 %	35.0 %	29.4 %
1	100.0 %	24.5 %	8.5 %
2	100.0 %	17.4 %	6.0 %
3	94.5 %	12.9 %	4.7 %
4	100.0 %	1.0 %	0.4 %
5	100.0 %	0.8 %	0.3 %
ECOGRAFIA	**77.3 %**	**94.9 %**	**28.8 %**
6	91.6 %	59.8 %	15.3 %
7	61.0 %	35.1 %	13.5 %
ECOC	**99.7 %**	**99.4 %**	**14.0 %**
8	99.7 %	99.4 %	14.0 %
DOPPLER	**100.0 %**	**67.3 %**	**8.0 %**
9	100.0 %	67.3 %	8.0 %

Table 5. Topic detection for dataset *Unina*: in this table it has been statistically described the distribution of document in the appropriate clusters, with reference to the macrocateories in exam. In the first column is reported the cluster ID; in the second group of columns is reported the statistical characterization of the document distribution into the categories of reference; in the third column is reported the cluster coverage; in the end are listed the most significant groups of features with the english translation in brackets.

Cluster	Rel. Att.	Amm.	Bandi	Didattica	Coverage	Top features
0	3 %		29 %	**67 %**	43.9 %	emanato, norme, seguito, conto, u.s.r *(emanated, standards, followed, account, r.e.d. (Regional Education Department)*
1	21 %	23 %	15 %	**41 %**	15.6 %	gara, foro, cauzione, possedute, addebito *(tender, forum, security deposit, owned, charging)*
2				**100 %**	6 %	facilitarne, aggrega, prefiggono, consecutivo, pianifica *(facilitate it, aggregates, aim, consecutive, plans)*
3				**100 %**	1.9 %	regolamentazioni, oligopolio, monopolio, microeconomia, macroeconomia *(regulations, oligopoly, monopoly, microeconomics, macroeconomics)*
4				**100 %**	1.5 %	opzione, roffredo, lictera, spagnole *(option, roffredo, lictera, spanish)*
5				**100 %**	1.1 %	pneumoconiosi, extraepatiche, ards, propriocettiva *(pneumoconiosis, extrahepatic, ards, proprioceptive)*
6				**100 %**	0.9 %	ril, discip, coordinatore, periimplantari, dental *(ril, discip, coordinator, peri-implant, dental)*
7				**100 %**	0.2 %	anafilotossine, passivazione, overjet, overbite, otori *(anaphylatoxins, passivation, overjet, overbite, otori)*
8	**99 %**			1 %	20.5 %	legittimata, odierna, giudizi, alfabetico, riunione *(legitimated, current, judgments, alphabetical, meeting)*
9			**100 %**		8.5 %	incombenza, destituzione, rimborsabile, risiedere, assunzione *(chore, destitution, refundeble, reside, recruitment)*

Table 6. Topic detection for dataset *Medical records*: in this table it has been statistically described the distribution of document in the appropriate clusters, with reference to the macrocateories in exam. In the first column is reported the cluster ID; in the second group of columns is reported the statistical characterization of the document distribution into the categories of reference; in the third column is reported the cluster coverage; in the end are listed the most significant groups of features with the english translation in brackets.

Cluster	Radiologia	Ecografia	Ecoc	Consulenze	Endoscopia	Doppler	Intervento	Coverage	Top features
0	41.0%	4.1%	0.3%	2.6%	15.3%	13.2%	23.5%	29.4%	norma, alterazioni, esame, eseguito, volume (*norm, alteration, exam, executed, volume*)
1	100.0%							8.5%	compatibile, cardiaca, lesioni, aortica, immagine, focolaio (*compatible, executed, lesion, aortic, image, locus*)
2	100.0%							6.0%	carattere, periferica, attivit , ilare, distribuzione (*character, peripheral, activity, hilar, distribution*)
3	94.5%			5.5%				4.7%	ingrandita, stasi, cardiaca, piccolo, costo (*overstated, stand-still, cardiac, small, cost*)
4	100.0%							0.4%	prominenza, attivit ,regolare, arco, periferica marginale (*prominence, activity, regular, frame, peripheral, marginal*)
5	100.0%							0.3%	aerea, bozzatura, tracheale, emidiaframma, scoliotica (*aerial, hump, tracheal, hemidiaphragm, scoliotic*)
6	8.4%	91.6%						15.3%	fegato, pancreas, milza, colecisti, reni (*liver, pancreas, spleen, Gallbladder, kidneys*)
7	12.0%	61.0%		6.2%	20.8%			13.5%	tiroide, lobo, prevalente, nodulari, componente (*thyroid, lobe, prevailing, nodular, component*)
8			99.7%	0.3%				14.0%	sezioni, normali, aortica, sinistre, deficit (*sections, normal, aortic, lefts, deficit*)
9						100.0%		8.0%	intimale, vasale, ispessimento, assi, succlavio (*intimal, vascular, thickening, axes, subclavian*)

cluster of documents or to use approaches to clean this data [15] or to use multi classification schemas [16].

Within the Tables 5 and 6 we represented the *top features* evaluated for each cluster with the percentage of documents belongs to each original category.

For the first corpus the most significant features are listed in Table 5.

As we can see, with clustering we obtained 10 groups fairly homogeneous with the exception of clusters 0 and 1, all others characterize well the contained documents. Moreover, we see that the selected features actually represent the topics. For instance:

- cluster 9 is identified by terms like *incombenza, destituzione, rimborsabile, risiedere,* and *assunzione* that are actually easy to find in documents belonging to the class *Bandi*, while they are more rare in the other categories;
- cluster 8 is represented by *legittimata, odierna, giudizi, alfabetico,* and *riunione,* which are frequent in meeting reports.

The most significant features extracted from medical reports are listed in Table 5. Also in this case, with clustering we have obtained 10 clusters, and with the exception of cluster 0 and cluster 7, all others groups are fairly homogeneous.

With regards to the topics, we can apply similar consideration as for previous case: the selected features actually represent the topics of the clusters. For instance:

- cluster 6 is represented by terms like *fegato, pancreas, milza, colecisti* and *reni* that actually are all organs for which you can do an ultrasonography;
- cluster 8 is identified by *sezioni, normali, aortica, sinistre* and *deficit,* which are terms related to the heart, frequent in reports about echocardiographies.

5   Conclusions

Topic extraction from documents is a challenging problem within the data mining field. The main motivation is due to its effectiveness in many tasks such as: information retrieval, information filtering and organization of documents collection in digital library.

In this paper we presented a methodology to implement an unsupervised topic detection for high dimensional datasets.

To this aim we used a preliminary clustering approach over the $tf - idf$ matrix computed starting from the corpora and we built n binary problems, one for each cluster obtained; we considered a supervised features selection over such problems to select the most important *features* and consequently the associated *topics.*

We showed the effectiveness of this approach on two different corpora, *UNINA* and *Medical Records,* obtaining interesting results.

As feature work we are planning to evaluate a set of distance measures to automatically figure out the *degree of belonging* between the *selected features* sets and the most interesting topics sets.

References

1. Wartena, C., Brussee, R.: Topic detection by clustering keywords. In: 19th International Workshop on Database and Expert Systems Application, DEXA 2008, pp. 54–58. IEEE (2008)
2. Jia Zhang, I., Madduri, R., Tan, W., Deichl, K., Alexander, J., Foster, I.: Toward semantics empowered biomedical web services. In: 2011 IEEE International Conference on Web Services (ICWS), pp. 371–378 (2011)
3. Bolelli, L., Ertekin, Ş., Giles, C.L.: Topic and trend detection in text collections using latent dirichlet allocation. In: Boughanem, M., Berrut, C., Mothe, J., Soule-Dupuy, C. (eds.) ECIR 2009. LNCS, vol. 5478, pp. 776–780. Springer, Heidelberg (2009)
4. Seo, Y.W., Sycara, K.: Text clustering for topic detection (2004)
5. Song, Y., Du, J., Hou, L.: A topic detection approach based on multi-level clustering. In: 2012 31st Chinese Control Conference (CCC), pp. 3834–3838. IEEE (2012)
6. Zhang, D., Li, S.: Topic detection based on k-means. In: 2011 International Conference on Electronics, Communications and Control (ICECC), pp. 2983–2985 (2011)
7. Manning, C.D., Schütze, H.: Foundations of Statistical Natural Language Processing. MIT Press, Cambridge (1999)
8. Amato, F., Gargiulo, F., Mazzeo, A., Romano, S., Sansone, C.: Combining syntactic and semantic vector space models in the health domain by using a clustering ensemble. In: HEALTHINF, pp. 382–385 (2013)
9. Amato, F., Mazzeo, A., Moscato, V., Picariello, A.: Semantic management of multimedia documents for e-government activity. In: International Conference on Complex, Intelligent and Software Intensive Systems, CISIS 2009, pp. 1193–1198. IEEE (2009)
10. Bellman, R.: Dynamic Programming. Princeton University Press, Princeton (1957)
11. Amato, F., Mazzeo, A., Penta, A., Picariello, A.: Knowledge representation and management for e-government documents. In: Mazzeo, A., Bellini, R., Motta, G. (eds.) E-Government ICT Professionalism and Competences Service Science, pp. 31–40. Springer, USA (2008)
12. Amato, F.M., Penta, A., Picariello, A.: Building RDF ontologies from semi-structured legal documents, complex, intelligent and software intensive systems. In: International Conference on CISIS 2008 (2008)
13. Holmes, G., Donkin, A., Witten, I.H.: Weka: a machine learning workbench. In: Proceedings of the 1994 Second Australian and New Zealand Conference on Intelligent Information Systems, pp. 357–361 (1994)
14. Pelleg, D., Moore, A.W.: X-means: extending k-means with efficient estimation of the number of clusters. In: Proceedings of the Seventeenth International Conference on Machine Learning, pp. 727–734. Morgan Kaufmann, San Francisco (2000)
15. Gargiulo, F., Sansone, C.: SOCIAL: self-organizing classifier ensemble for adversarial learning. In: El Gayar, N., Kittler, J., Roli, F. (eds.) MCS 2010. LNCS, vol. 5997, pp. 84–93. Springer, Heidelberg (2010)
16. Gargiulo, F., Mazzariello, C., Sansone, C.: Multiple classifier systems: theory, applications and tools. In: Bianchini, M., Maggini, M., Jain, L.C. (eds.) Handbook on Neural Information Processing. ISRL, vol. 49, pp. 335–378. Springer, Heidelberg (2013)

A Framework for Real-Time Evaluation of Medical Doctors' Performances While Using a Cricothyrotomy Simulator

Daniela D'Auria and Fabio Persia[✉]

Dip. di Ingegneria Elettrica e Tecnologie dell'Informazione,
University of Naples "Federico II", Naples, Italy
{daniela.dauria4,fabio.persia}@unina.it

Abstract. Cricothyrotomy is a life-saving procedure performed when an airway cannot be established through less invasive techniques. One of the main challenges of the research community in this area consists in designing and building a low-cost simulator that teaches essential anatomy, and providing a method of data collection for performance evaluation and guided instruction as well.

In this paper, we present a framework designed and developed for activity detection in the medical context. More in details, it first acquires data in real time from a cricothyrotomy simulator, when used by medical doctors, then it stores the acquired data into a scientific database and finally it exploits an *Activity Detection Engine* for finding expected activities, in order to evaluate the medical doctors' performances in real time, that is very essential for this kind of applications. In fact, an incorrect use of the simulator promptly detected can save the patient's life. The conducted experiments using real data show the approach efficiency and effectiveness. Eventually, we also received positive feedbacks by the medical personnel who used our prototype.

Keywords: Activity detection · Scientific databases · Cricothyrotomy simulator · Medical simulator

1 Introduction

Robotic surgery, computer-assisted surgery, and robotically-assisted surgery are terms for technological developments that use robotic systems to aid in surgical procedures. Robotically-assisted surgery was developed to overcome the limitations of minimally-invasive surgery and to enhance the capabilities of surgeons performing open surgery.

In the case of robotically-assisted minimally-invasive surgery, instead of directly moving the instruments, the surgeon uses one of five methods to control the instruments; either a direct telemanipulator or through computer control. A telemanipulator is a remote manipulator that allows the surgeon to perform the normal movements associated with the surgery whilst the robotic arms carry

© Springer International Publishing Switzerland 2015
M. Helfert et al. (Eds.): DATA 2014, CCIS 178, pp. 182–198, 2015.
DOI: 10.1007/978-3-319-25936-9_12

out those movements using end-effectors and manipulators to perform the actual surgery on the patient. In computer-controlled systems the surgeon uses a computer to control the robotic arms and its end-effectors, though these systems can also still use telemanipulators for their input. One advantage of using the computerised method is that the surgeon does not have to be present, but can be anywhere in the world, leading to the possibility for remote surgery.

In the case of enhanced open surgery, autonomous instruments (in familiar configurations) replace traditional steel tools, performing certain actions (such as rib spreading) with much smoother, feedback-controlled motions than could be achieved by a human hand. The main object of such smart instruments is to reduce or eliminate the tissue trauma traditionally associated with open surgery without requiring more than a few minutes' training on the part of surgeons. This approach seeks to improve open surgeries, particularly cardio-thoracic, that have so far not benefited from minimally-invasive techniques.

In this paper, we present a framework designed and developed for activity detection in the medical context. The context of use is very concrete and relevant in the robotic surgery research area, as it is represented by a *cricothyrotomy simulator* built by the BioRobotics Laboratory of the University of Washington, Seattle (USA) [15–17]. Such a simulator is very useful for helping both patients and medical doctors when a cricothyrotomy procedure is performed. Our main aim consists in making the medical doctors able to get a real-time feedback about their performances when using the simulator, that is very essential for this kind of applications. Moreover, this real-time feedback can even save the patient's life, as in this way a serious error of the medical doctor during the procedure could be fixed by an immediate recovery procedure. In order to reach this goal, our framework first acquires data in real time from the simulator, when used by medical doctors; then, it stores the acquired data into a *scientific database* and uses an *Activity Detection Engine* which exploits a specific *knowledge base* and apposite *algorithms* for finding *expected activities*, corresponding to the performances obtained by the medical doctors when using the simulator. Thus, we model the so-called expected activities with *stochastic automata* [12,13] and exploit the activity detection algorithms presented by [14], which are based on temporal information, that is clearly essential in such a context.

In order to best clarify what is the main goal of this work and our vision of *expected activities*, in what follows we will briefly describe a simple and usual example from a real case study. Let us informally consider an *expected activity* as an ordered sequence of actions describing the interactions of a medical doctor with the simulator. An example of such a sequence in a simulator log could be: ⟨Start Procedure, Pad C touched with SCALPEL, Pad C touched with KELLY FORCEPS, Pad C touched with HOOK, Pad C released, Pad B touched with KELLY FORCEPS, End Procedure⟩ (see Fig. 1). For the sake of simplicity, let us consider the graph in Fig. 2 as the only model representing our knowledge base, corresponding to an incorrect use of the simulator. In this example, we are overlooking the temporal constraints fixed between an action and the following one within the activity model and also the transition probabilities specified on

Start procedure	Pad C touched with SCALPEL	Pad C touched with KELLY FORCEPS	Pad C touched with HOOK	Pad C released	Pad B touched with KELLY FORCEPS	End Procedure

Fig. 1. Example of log sequence.

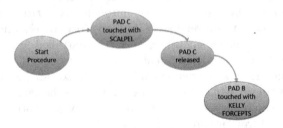

Fig. 2. Model of an expected (bad) activity.

each edge. Further in the paper we will discuss how conveniently modelled time-bounds can be used to more accurately detect expected activity occurrences.

At this point, we have to specify that the procedure for discovering expected activities follows an *event-driven* approach and is thus invoked every time that a new action is recorded by our system. In this example, the procedure is called for the first time when the *Start Procedure* action is detected - no occurrences of the only activity model is discovered - for the second time when the *Pad C touched with SCALPEL* action is detected - no occurrences found again - and so on. Then, when the *Pad B touched with KELLY FORCEPS* action is recorded, if all temporal constraints (that we are overlooking in this example) are satisfied, an occurrence of the expected activity model is found, then an alert is generated thus stopping the procedure, in order to immediately start an appropriate recovery process. This is a classic example of how important the acquisition of a real-time feedback from the simulator is: an incorrect use promptly detected can save the patient's life.

The paper is organized as in the following. Section 2 reports a brief overview of the methods used in the literature for exploiting temporal information in different contexts. Section 3 briefly describes the context of use of our prototype architecture, which is the *cricothyrotomy simulator* designed by the University of Washington, Seattle. Section 4 shows the model used for activity detection, while Sect. 5 describes the architecture of the overall framework. Section 6 presents the experiments conducted in a real medical context. Eventually, Sect. 7 discusses some conclusions and future work.

2   Related Work

Reasoning techniques are very essential in many application domains, such as video surveillance, cyber security, fault detection, fraud detection and in clinical domain, as well. In all cases, *temporal information* is crucial. For instance, for what the clinical research concerns, investigating disease progression is practical only by definition of a time line; otherwise, possible causes of a clinical condition have to be found by referring to a patient's past clinical history. In [2], the basic concepts of temporal representation in the medical domain have been described in order to include: category of time (natural, conventional, logical), structure of time (line, branch, circular, parallel), instant of time vs. interval, and, absolute time vs. relative time. Anyway, this is still a challenging and active subject of research. The main goal of [1] consists in creating a special purpose query language for clinical data analytics (CliniDAL) to place in any clinical information system (CIS) and answer any answerable question from the CIS. More in details, a category scheme of five classes of increasing complexity, including point-of-care retrieval queries, descriptive statistics, statistical hypothesis testing, complex hypotheses of scientific studies and semantic record retrieval have been designed to capture the scope encompassed by CliniDAL's objectives [3]. However, a review of temporal query languages reflects that the importance of time has led to the development of custom temporal management solutions, which are mostly built to extend relational database systems (for instance, T4SQL [4]). Many efforts in the relational database field have been conducted for developing expressive temporal query languages; nevertheless, they still suffer from two issues: firstly, they are only applicable to structural relational databases; secondly, it is difficult for hospital staff with poor IT skills to apply them. On the other hand, in most ontology based approaches composing queries can be difficult due to a complex underlying model representation and lack of expressivity.

In other contexts, such as video surveillance, cyber security and fault detection, the reasoning techniques using temporal information are broadly used for *activity detection*. Thus, several researchers have studied how to search for specifically defined patterns of normal/abnormal activities [6]. Vaswani et al. [7] study how HMMs can be used to recognize complex activities, while Brand et al. [8] and Oliver et al. [9] use coupled HMMs. Hamid et al. [10] use Dynamic Bayesian networks (DBNs) to capture causal relationships between observations and hidden states. Albanese et al. [5] developed a stochastic automaton-based language to detect activities in video, while Cuntoor et al. [11] presented an HMM-based algorithm. In contrast, [12,13] start with a set A of *activity models* (corresponding to innocuous/dangerous activities) and find observation sequences that are not sufficiently explained by the models in A. Such unexplained sequences reflect activity occurrences that differ from the application's expectations.

Other relevant works exploiting an *events sequence* definition are [20,21]: in particular, [20] automatically identify cleansing activities, namely a sequence of actions able to cleanse a dirty dataset, which today are often developed manually by domain-experts, while [21] describe how a model based cleansing framework is extended to address integration activities as well.

3   Context of Use: A Cricothyrotomy Simulator

Modern airway protocols involve many techniques to restore ventilation including bag-mask-ventilation, placement of a laryngeal mask airway, and intubation with or without videolaryngoscope. In cases where conservative measures fail or when contraindicated, the only methods remaining to re-establish ventilation may be surgical. In the developing world where devices such as the videolaryngoscope may not be available, accurate knowledge and training in the creation of a surgical airway may have a significant impact on patient outcomes.

A *cricothyrotomy* is a life-saving procedure performed when an airway cannot be established through less invasive techniques: although performing such a procedure seems relatively straightforward, studies have shown that those performed in the pre-hospital setting were mostly unsuccessful [22]. A review of 54 emergency cricothyrotomies found that the majority of the procedures performed in the field were unsuccessful or resulted in complications [23]. A military team identified gap areas in the training of cricothyrotomy in emergency situations; these included lack of anatomical knowledge including *hands on* palpation exercises, poor anatomy in medical mannequins, and non-standard techniques [24].

Most of the unsuccessful attempts were due to inaccurate placement, and incorrectly identifying anatomy. If the anatomy is not properly identified, it is unlikely that the procedure will be successful. Further, a large review of emergency airway cases found that emergency cricothyrotomies performed by anesthesiologists were successful in only 36 % of instances [25]. Although many reports suggest that the success rate of surgical airway placement is low, publications from advanced centers with extensive training for airway protocols including simulation show that pre-hospital cricothyrotomy success rates can be as high as 91 % [26]. Studies such as this suggest that with adequate training, the success rate of cricothyrotomy can be dramatically improved. Thus, an improved method of training needs to be provided for this rare, but life-saving procedure.

For such reasons, the BioRobotics Laboratory of the University of Washington, Seattle (USA) developed a low-cost cricothyrotomy simulator [15–17] from readily available components that is equipped with inexpensive sensors. The simulator emphasizes the palpation and the correct identification of anterior cervical anatomy and has the ability to record in real time the contact location of instruments on the trachea model during the full duration of the simulated procedure.

3.1   Simulator Design

The trachea model is disposable and is replaced after each procedure. To minimize costs, the trachea is made of cardboard with fixed size and dimension according to the trachea of an average adult. Foam strips were cut for the cartilaginous tracheal rings and were attached on the trachea with appropriate spacing. The *thyroid* and *cricoid* cartilages are permanent parts made of ABS plastic, using a 3D printer. These components are fixed onto a wooden base, and firmly support the trachea model. Conductive foils are used as low-cost sensors

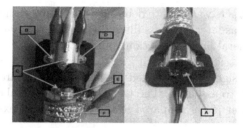

Fig. 3. Critical landmarks on trachea are covered with conductive foils (A: Posterior tracheal wall, B: Right lateral trachea and cricothyroid membrane, C: Midline cricothyroid membrane (correct placement of incision), D: Left lateral trachea and cricothyroid membrane, E: Cricoid cartilage, and F: Cartilaginous ring of lower tracheal wall).

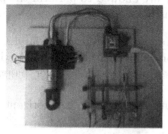 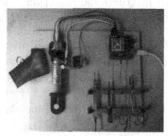

(a) *Trachea model is covered with inner bicycle tube as human skin.*   (b) *Six different landmarks represented by conductive foils and tools (scalpel, hook and forceps) are connected to microcontroller for data collection.*

Fig. 4. Low-cost cricothyrotomy simulator.

to detect six critical landmarks (identified by A-F letters) that providers might contact during the procedure (Fig. 3). The conductive foils cover landmarks on the trachea model. Only one of these six landmarks, the cricothyroid membrane itself, is the correct area to contact and make an opening. Other landmarks like the posterior tracheal wall and lateral locations into the tracheoesphageal grooves should be avoided during the procedure.

The model is fitted with an *Arduino Uno* microcontroller board based on the Atmel Atmega 328 microprocessor with a mounted 8 × 8 LED matrix-based display for user interface capability. The microcontroller records the contact data of the instruments (scalpel, tracheal hook, and hemostat) onto the six conductive foils, as each instrument is wired. During the procedure, when a closed circuit is detected between the instrument and a patch of foil, the event is recorded and labeled with the time in milliseconds. Breaking contact is similarly recorded. Traditional matrix scanning techniques are used by the microcontroller to detect connections between the instruments and the foil patches. A minimum time

(20 ms) between contacts was used to debounce the inputs to the microcontroller. The resulting data was later low-pass filtered with a cutoff frequency of 10 Hz in accordance with general human reaction time.

The simulator's design is shown in Fig. 4.

Moreover, the design was optimized for materials that are low-cost, widely available and simple to assemble. The total cost of the simulator was less than $50, which has a lower price compared to the ones of other existing commercial simulators.

3.2   How to Use the Simulator

Medical doctors who want to use the simulator are firstly forced to watch a video tutorial published by the New England Journal of Medicine [18,19]. After watching the instructional materials, they are allowed to perform the procedure on the simulator following the instructions below:

- Step 1: Palpate the cricothyroid membrane. Immobilize the larynx with the non-dominant hand and perform the procedure with the dominant hand.
- Step 2: Incise the skin (bicycle inner tube) vertically after palpating the cricothyroid membrane.
- Step 3: Incise the cricothyroid membrane on trachea model horizontally (1 cm length).
- Step 4: Insert the tracheal hook into cricoid cartilage.
- Step 5: Insert the hemostat and to expand the airway opening vertically and horizontally.
- Step 6: Insert the endotracheal tube.

As these steps were performed, all the data (contact locations on trachea model, instruments information, contact duration and total time) were recorded by the microcontroller; procedures were also video-recorded for analysis.

Thus, after the description of the used *cricothyrotomy simulator*, the importance of temporal information for defining apposite *activity detection algorithms* is definitely clear. The following sections describe how our framework has been designed and developed for helping both medical doctors and patients when a cricothyrotomy is performed.

4   Modeling Expected Activities

This section describes the model that we have defined in order to derive a formal definition of *Expected Activity* for medical context. We use the temporal probabilistic graph proposed by [12,13], so that the elapsed time between observations also plays a role in defining whether a sequence of observations belongs to an activity, differently from what happens in other models, such as Hidden Markov Chains. We assume the existence of a finite set S of *action symbols*, corresponding to atomic events that can be detected by the *Arduino microcontroller board*, as described in Sect. 3.

4.1 Basic Definitions

An *Expected Activity* is a labeled directed graph A= (V ,E, δ, ρ) where: (i) V is a finite set of nodes labeled with action symbols from S; (ii) $E \subseteq V \times V$ is a set of edges; (iii) $\delta : E \to \mathbb{N}^+$ associates with each edge $\langle v_i, v_j \rangle$ an upper bound of time that can elapse between v_i and v_j; (iv) $\rho : E \to (0,1)$ is a function that associates a probability distribution with the outgoing edges of each node, i.e. $\forall v \in V \sum_{\langle v,v' \rangle \in E} \delta(\langle v, v' \rangle) = 1$; (v) there exists an initial node I in the activity definition, i.e. $\{v \in V \mid \nexists\ v' \in V \text{ s.t. } \langle v', v \rangle \in E\} \neq \emptyset$; (vi) there exists a final node F in the activity definition, i.e. $\{v \in V \mid \nexists\ v' \in V \text{ s.t. } \langle v, v' \rangle \in E\} \neq \emptyset$.

We assume the existence of a finite set S of *action symbols* representing particular interactions (for instance, *Pad C touched with SCALPEL, PAD C released*) between medical doctors and the simulator. Figure 5 shows an expected activity model representing simple interactions between a medical doctor and the cricothyrotomy simulator.

Then, we define an *instance* of an expected activity as a specific path in A from the initial node to the end node.

An *instance* of an *Expected Activity* (V, E, δ, ρ) is a finite sequence $\langle v_1, ..., v_m \rangle$ of nodes in V such that: (i) $\langle v_i, v_{i+1} \rangle \in E$ for $1 < i < m$; (ii) $\{v \mid \langle v, v_1 \rangle \in E\} = 0$, i.e. v_1 is the start node I; (iii) $\{v \mid \langle v_m, v \rangle \in E\} = 0$, i.e. v_m is the final node F. The probability of the instance is $\prod_{i=1}^{m-1} \rho(\langle v_i, v_{i+1} \rangle)$.

We work with sequences of time-stamped events. Let us assume that the number of observable events in our domain is finite, each event can then be associated to a different action symbol in the set S. We define an *observed event* as a pair $\omega = (s, ts)$, where $\omega.s$ is the action symbol associated to the event and $\omega.ts$ is the time stamp at which s was observed.

We call a *Simulator Log* Ω a finite sequence of log entries ω_i.

Now, we are in the position of defining the concept of *Activity Occurrence*.

Let Ω be a *Simulator Log* and A=(V, E, δ, ρ) an Expected Activity. An *occurrence* of A in Ω is a sequence $\langle (\omega_1, v_1)...(\omega_m, v_m) \rangle$ where: (i) $\langle \omega_1, ..., \omega_m \rangle$

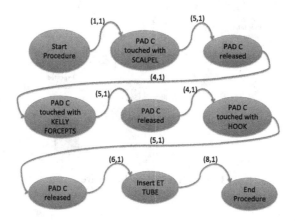

Fig. 5. An example of Expected Activity Model.

is a subsequence of Ω such as $\omega_i = (\omega_i.ts, \omega_i.s)$, $\omega.s$ being an action symbol from S and $\omega.ts$ the associated time-stamp; (ii) $\langle v_1, ..., v_m \rangle$ is an instance of A; (iii) $v_i = \omega_i.s$ for $1 < i < m^1$; (iv) $\omega_{i+1}.ts - \omega_i.ts \leq \delta(\langle v_i, v_{i+1} \rangle)$ for $1 < i < m$.

The probability $p(o)$ of the occurrence o should be the probability of the instance $\langle v_1, ..., v_m \rangle$. Of course, shorter activities usually have higher probabilities. Therefore, since we compare occurrences of different activity models despite their different lengths, we introduce the relative probability $p^*(o) = p(o)/p(max)$. When computing $p^*(o)$ for a given occurrence o, we consider $p(max)$ as the highest probability of any instance of A when ignoring each instance's self-loops.

Thus, once we have given the previous formal definitions for defining our *expected activity model*, we can describe in Sect. 5 the proposed architecture for real-time evaluation of medical doctors' performances when using a cricothyrotomy simulator.

5   The Proposed Architecture

The theoretical model has been exploited to develop a framework for the detection of expected activities in medical context scientific databases. The structure of the system is based on a modular architecture, as shown in Fig. 6, which allows the medical doctors to get a real-time feedback about their performances when using the simulator.

The following subsections describe the single components of the overall system architecture.

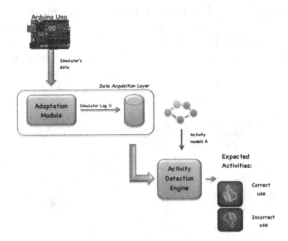

Fig. 6. System architecture.

---

1 v_i refers both to the node v_i in A and the action symbol s_i labeling it.

5.1   The Arduino Microcontroller Board

As also mentioned in Sect. 3, the *Arduino microcontroller board* allows us to capture in real time the contact data of the instruments (scalpel, tracheal hook, and hemostat) from six different landmarks of the simulator. In such a way, this component records the series of time-stamped events, corresponding to the medical doctors' interactions with the simulator. More in details, events are defined as the start and end times of contacts between specific instruments and surfaces on the anatomical model. Other types of events are defined in terms of readings from different sensor types. Thus, events are represented by a series of symbols (ASCII characters).

Data are encoded as follows:

- The first single digit number indicates the instrument (1 means *Scalpel*, 2 *Hemostat* and 3 *Tracheal Hook*).
- The character indicates which foil patch is touched: upper-case for making contact and lower-case for breaking contact. More in details, A means *Posterior tracheal wall*, B the *Right lateral trachea and cricothyroid membrane*, C the *Midline cricothyroid membrane (correct placement of incision)*, D the *Left lateral trachea and cricothyroid membrane*, E the *Cricoid cartilage* and F the *Cartilaginous ring of lower tracheal wall*.
- The last number is the time in milliseconds.

Then, the data captured in this way represent the input of the *Data Acquisition* component.

5.2   The Data Acquisition Component

The *Data Acquisition* component includes an *Adaptation Module* that converts the data captured using the *Arduino* in a format suitable to the *Activity Detection Engine* (i.e. the *Simulator Log*): it also saves them into a scientific database, which is also able to store personal information about the medical doctors who are using the simulator.

5.3   The Activity Detection Engine

The *Activity Detection Engine* takes as inputs time-stamped user data collected in the *Simulator Log* and a set of activity models representing our *knowledge base* to find the activity occurrences matching such models. These models have been previously defined by domain experts who have classified them in two different categories: the *good activities*, corresponding to a correct use of the simulator and the *bad activities*, corresponding to an incorrect use of the simulator. Figures 7 and 8 show two model examples of a *good activity* (Fig. 7), corresponding to an excellent performance of the medical doctor and a *bad activity* (Fig. 8), corresponding to a very bad performance.

Expected activity occurrences in a data stream are efficiently detected using *tMagic* [14], which allows to solve the problem of finding occurrences of high-level

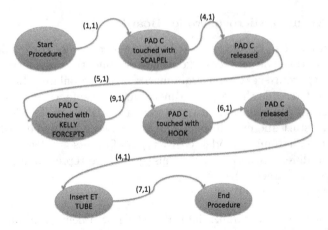

Fig. 7. Model of an excellent performance.

activity model in an observed data stream. As a matter of fact, they propose a data structure called *temporal multiactivity graph* to store multiple activities that need to be concurrently monitored, corresponding to our knowledge base of *good* and *bad* activities. They then define an index called *Temporal Multi-activity Graph Index Creation (tMAGIC)* that, based on this data structure, examines and links observations as they occur. Finally, they define an algorithm to solve the *evidence problem* that tries to find all occurrences of an activity (with probability over a threshold) within a given sequence of observations.

The procedure for the identification of the *Expected Activities* follows an *event-driven approach*, as it is automatically invoked every time that a new action symbol is captured by the *Arduino microcontroller board*: in this way, if the *Activity Detection Engine* discovers a *bad activity* when the procedure

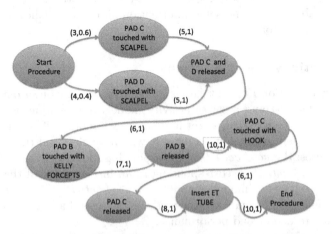

Fig. 8. Model of a bad performance.

has not been completed yet, it generates an alert inducing the medical doctor to immediately stop his procedure and to start the recovery process for saving the patient's life. In this way, we are able to evaluate the medical doctors' performances while using the simulator both during and after the procedure. Obviously, the *bad performances* identified during the procedure are definitely more dangerous than the ones detected after the procedure: in fact, the former ones could even cause the death of the patient and thus need to be stopped, while the latter ones are mostly slow or inaccurate procedures.

6 Experimental Results

This section shows the experimental evaluation of our framework. We present the experimental protocol for evaluating our framework in terms of *execution time scalability*, *detection accuracy* and *user satisfaction*. Eventually, we show the medical doctors' performances detected by our framework in a specific case study.

6.1 Evaluating Execution Time

We decided to measure[2] the execution time of our framework for detecting *Expected Activities* in the worst case (at most one action symbol for each millisecond) when varying the length of the Simulator Log and using the previously defined set of known activity models. More in details, the maximal length of Simulator Log considered has been 5 min, since a longer procedure would cause the death of the patient. The time for acquiring data using the *Arduino microcontroller* can be considered as negligible. Thus, the *Total Running Time* is given by the sum of the *Storing and Index Building Time* (higher value) and the *Query Time* (lower value), as shown in Fig. 9. However, the obtained *Total Running Time* can be considered very low even if we are considering the worst case.

6.2 Accuracy Results

100 medical doctors participated in a trial and used the simulator with our additional framework. The *Precision* and *Recall* metrics have been used to compute the accuracy [12,13], by comparing the Expected Activities discovered by our framework with a ground truth defined by experts who watched the recordings of the medical doctors' performances several times.

We use $\{A_i^a\}_{i \in [1,m]}$ to denote the Expected Activities returned by our framework and $\{A_j^h\}_{j \in [1,n]}$ to denote the activities flagged as expected by human annotators. Precision and recall were computed as follows:

---

[2] All experiments presented in this Section were conducted on a machine running Mac OS X 10.9.1, and mounting a 2 GHz *Intel Core i7* processor with a 8 GB, 1600 MHz *DDR3*.

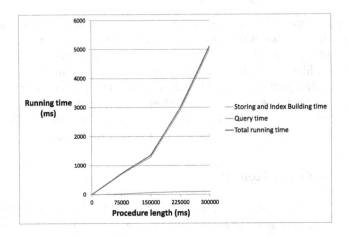

Fig. 9. Framework running times.

$$P = \frac{|\{A_i^a | \exists A_j^h \ s.t. \ A_i^a = A_j^h\}|}{m} \qquad (1)$$

and

$$R = \frac{|\{A_j^h | \exists A_i^a \ s.t. \ A_i^a = A_j^h\}|}{n} \qquad (2)$$

We achieved an *average Precision* of *81 %* and an *average Recall* of *98 %*, that can be considered a very encouraging result. Analyzing the results in detail, we figured out that errors were due to the incompleteness of the knowledge base, that can be extended all the time.

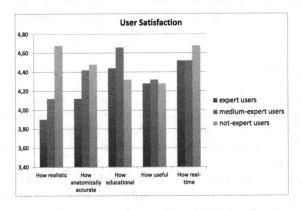

Fig. 10. User satisfaction.

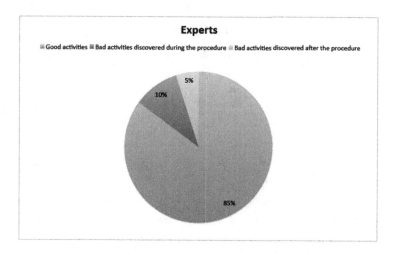

Fig. 11. Expert users' performance.

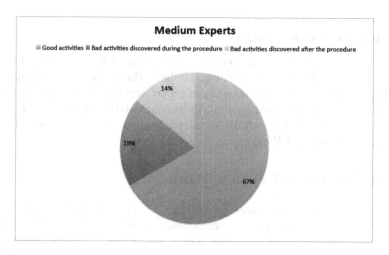

Fig. 12. Medium-expert users' performance.

6.3 User Satisfaction

After completing the procedure, the medical doctors, classified in three different categories (expert, medium-expert and not expert users), filled out a questionnaire to report their level of training, experience and their impressions of the simulator. Each subject was asked to answer 5 questions about the simulator (How realistic, How anatomically accurate, How educational, How useful, How real-time) using a 5-point Likert scale. As we can see in Fig. 10, subjects (especially not-expert users) expressed positive opinions about their experiences with the simulator.

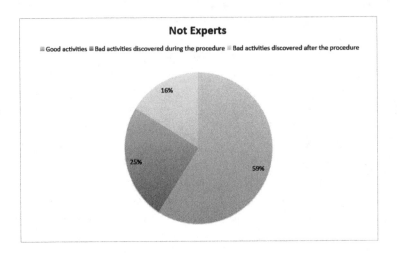

Fig. 13. Not-expert users' performance.

6.4 Medical Doctors' Performances

In this subsection, we show the performances obtained by 100 medical doctors within our case study. In particular, Fig. 11 shows the performances of the *expert users*, Fig. 12 the performances of *medium-expert users* and Fig. 13 the *not-expert users*. Therefore, the obtained results can be considered very good, thus proving the relevance of the *cricothyrotomy simulator* built by the BioRobotics Laboratory of the University of Washington in Seattle.

7   Conclusions and Future Work

This work presented a framework for activity detection in the medical context. It starts acquiring data from a *cricothyrotomy simulator*, when used by medical doctors and then it stores the captured data into a *scientific database*. Finally, it exploits some stable activity detection algorithms for discovering expected activities, corresponding to specific performances obtained by the medical doctors when using the simulator, that can be detected both during and after the procedure. The experiments showed encouraging results concerning *efficiency*, *effectiveness* and *user satisfaction*.

Future work will be devoted to enlarge our experimentation and to plan to integrate the prototype in more complex and thorny applications by adding new functionalities and, if necessary, additional layers to the overall system architecture. For example, a potential application of this tool could consist in detecting potential safety hazard *in advance*, for instance, using machine learning techniques and observations learned during the training of medical personnel, or even in suggesting the correct recovery process to apply when a bad activity is discovered during the procedure. Moreover, data mining techniques could be used in an offline setting to analyze in detail the medical doctors' performances.

References

1. Safari, L., Patrick, J.D.: A temporal model for clinical data analytics language. In: 35th Annual International Conference of the IEEE, Engineering in Medicine and Biology Society (EMBC), pp. 3218–3221 (2013)
2. Zhou, L., Hripcsak, G.: Temporal reasoning with medical data–a review with emphasis on medical natural language processing. J. Biomed. Inform. **40**, 183–203 (2007)
3. Patrick, J.D., Safari, L., Cheng, Y.: Knowledge discovery and knowledge reuse in clinical information systems. In: Proceedings of the 10th IASTED International Conference on Biomedical Engineering (BioMed 2013), Innsbruck, Austria (2013)
4. Combi, C., Montanari, A., Pozzi, G.: The T4SQL temporal query language. In: The Sixteenth ACM Conference on Information and Knowledge Management, pp. 193–202. ACM, Lisbon (2007)
5. Albanese, M., Moscato, V., Picariello, A., Subrahmanian, V.S., Udrea, O.: Detecting stochastically scheduled activities in video. In: Proceedings of the 20th International Joint Conference on Artificial Intelligence, pp. 1802–1807 (2007)
6. Hongeng, S., Nevatia, R.: Multi-agent event recognition. In: Proceedings of the International Conference on Computer Vision (ICCV), pp. 84–93 (2001)
7. Vaswani, N., Chowdhury, A.K.R., Chellappa, R.: Shape activity: a continuous-state HMM for moving/deforming shapes with application to abnormal activity detection. IEEE Trans. Image Process. **14**(10), 1603–1616 (2005)
8. Brand, M., Oliver, N., Pentland, A.: Coupled hidden Markov models for complex action recognition. In: Proceedings of the IEEECS Conference on Computer Vision and Pattern Recognition (CVPR), pp. 994–999 (1997)
9. Oliver, N., Horvitz, E., Garg, A.: Layered representations for human activity recognition. In: Proceedings of the IEEE Fourth Internaional Conference on Multimodal Interfaces (ICMI), pp. 3–8 (2002)
10. Hamid, R., Huang, Y., Essa, I.: Argmode - activity recognition using graphical models. In: Proceedings of the Computer Vision and Pattern Recognition Workshop (CVPRW), pp. 38–43 (2003)
11. Cuntoor, N.P., Yegnanarayana, B., Chellappa, R.: Activity modeling using event probability sequences. IEEE Trans. Image Process. **17**(4), 594–607 (2008)
12. Albanese, M., Molinaro, C., Persia, F., Picariello, A., Subrahmanian, V.S.: Discovering the top-k unexplained sequences in time-stamped observation data. IEEE Trans. Knowl. Data Eng. (TKDE) **26**(3), 577–594 (2014)
13. Albanese, M., Molinaro, C., Persia, F., Picariello, A., Subrahmanian, V.S.: Finding unexplained activities in video. In: International Joint Conference on Artificial Intelligence (IJCAI), pp. 1628–1634 (2011)
14. Albanese, M., Pugliese, A., Subrahmanian, V.S.: Fast activity detection: indexing for temporal stochastic automaton based activity models. IEEE Trans. Knowl. Data Eng. (TKDE) **25**(2), 360–373 (2013)
15. White, L., Bly, R., D'Auria, D., Aghdasi, N., Bartell, P., Cheng, L., Hannaford, B.: Cricothyrotomy simulator with computational skill assessment for procedural skill training in the developing world. J. Otolaryngol. - Head Neck Surg. **149**, p. 60 (2014)
16. White, L., Bly, R., D'Auria, D., Aghdasi, N., Bartell, P., Cheng, L., Hannaford, B.: Cricothyrotomy Simulator with Computational Skill Assessment for Procedural Skill Training in the Developing World, AAO-HNSF Annual Meeting and OTO Expo, Vancouver, BC, September 2013

17. White, L., D'Auria, D., Bly, R., Bartell, P., Aghdasi, N., Jones, C., Hannaford, B.: Cricothyrotomy simulator training for the developing word. In: 2012 IEEE Global Humanitarian Technology, Seattle, WA, October 2012

18. James Hsiao, M.D., Victor Pacheco-Fowler, M.D.: "Cricothyrotomy". N. Engl. J. Med. **358**, e25 (2008). doi:10.1056/NEJMvcm0706755. http://www.nejm.org/doi/full/10.1056/NEJMvcm0706755

19. BioRobotics Laboratory, University of Washington, Seattle, "Global Simulation Training in Healthcare". http://brl.ee.washington.edu/laboratory/node/2768

20. Boselli, R., Cesarini, M., Mercorio, F., Mezzanzanica, M.: Planning meets data cleansing. In: Proceedings of the Twenty-Fourth International Conference on Automated Planning and Scheduling (2014)

21. Boselli, R., Cesarini, M., Mercorio, F., Mezzanzanica, M.: A policy-based cleansing and integration framework for labour and healthcare data. In: Holzinger, A., Jurisica, I. (eds.) Interactive Knowledge Discovery and Data Mining in Biomedical Informatics. LNCS, vol. 8401, pp. 141–168. Springer, Heidelberg (2014)

22. Wang, H., Mann, N., Mears, G., Jacobson, K., Yealy, D.: Out-of-hospital airway management in the United States. Resuscitation **82**(4), 378–385 (2011)

23. King, D., Ogilvie, M., Michailidou, M., Velmahos, G., Alam, H., deMoya, M., Fikry, K.: Fifty-four emergent cricothyroidotomies: are surgeons reluctant teachers? Scand. J. Surg. **101**(1), 13–15 (2012)

24. Bennett, B., Cailteux-Zevallos, B., Kotora, J.: Cricothyroidotomy bottom-up training review: battlefield lessons learned. Mil. Med. **176**(11), 1311–1319 (2011)

25. Cook, T., Woodall, N., Frerk, C.: Major complications of airway management in the UK: results of the Fourth National Audit Project of the Royal College of Anaesthetists and the Difficult Airway Society. Part 1: Anaesthesia. Br. J. Anaesth. **106**(5), 617–631 (2011). Fourth National Audit Project

26. Warner, K., Sharar, S., Copass, M., Bulger, E.: Prehospital management of the difficult airway: a prospective cohort study. J. Emerg. Med. **36**(3), 257–265 (2009)

Author Index

Printed in the United States
By Bookmasters